The Drone Chronicles 2001 – 2016 Journal

ÉDITIONS CENTRE DE LA PHOTO— GRAPHIE GENÈVE

Spector Books

Preface
Joerg Bader

Rob van Leijsen's *The Drone Chronicles* is an artistic project with political ambition, on the contrary to what different juries may have thought, seeing a political activist project without artistic interest. The artistic aspect does not lie in the excellent and most neutral form that the graphic designer was able to give to his project, that primarely resembles a retail catalogue. The artistic value lies much more in the very gesture, of absurd appearance, of making an inventory of everything that could have been produced in the capitalist world in the realm of unmanned flying objects between 2001 and 2016. The affair becomes political when you have in front of your eyes all these military-style devices, which constitute 30% of all drones manufactured worldwide in the time period determined by the author.

UAVs are already part of the arsenal to control us and they are terribly feared by civilians in war zones. Apart from civilian jobs such as delivering pizzas or disrupting an airport for an entire day, military drone activity is roughly divided into two categories. Either the drones are loaded with missiles or they have built-in cameras to spy while on reconnaissance missions.

The Centre de la photographie Genève has presented in 2016 the exhibition *caméra(auto) control*, which focused on how we control ourselves (social control by selfies, for example) and how we are controlled by cameras. Among the 63 artists or collectives exhibiting, artists such as Sean Snyder or Adrien Missika designed their works with drones, Sébastien Reuzé documented a drone base in Indian Springs, near Las Vegas, Clément Lambelet reused images of aerial executions transmitted by drones and James Bridle had a drone shadow painted on Plainpalais, a large public square in Geneva. This book can therefore be seen as an extension of the exhibition launched three years ago.

All my gratitude goes to Rob van Leijsen for his perseverance, among other things, and and the sense of humour he has kept during the two years of preparation as well as to the entire team at Centre de la photographie Genève.

Introduction: I spy with my little eye something that begins with the letter D

Rob van Leijsen

Why are drones — or unmanned aerial vehicles (UAVs) — so present in our daily news feed and local retail stores, but do we — despite sky-high sales numbers — hardly see them fly in our public airspace? This paradox of invisible presence lies at the basis of *The Drone Chronicles*, a diptych that maps the evolution and integration of drones in our society from 2001 onwards. As a drone surveilling a battlefield or crowd, the book offers a contemporary view on drones without emphasising a particular category or usage. It discusses how this relatively new piece of technology has shifted from a commonly known aerial bombing machine to a multi-purpose and accessible tool.

The earliest example of an unmanned airplane made its appearance during WWI. The *Kettering Bug*, developed by Charles F. Kettering and Orville Wright for the US Army in 1918, was designed as an "aerial torpedo", or "flying bomb". Made out of wood and card-board, it had a wingspan of 4.5 meters and could carry an 82 kilograms bomb[1]. It was not until the end of WWII that the benefits of pilot-less, autonomous fighting planes came to full potential with the development of the *TDR1*[2], deployed by the US Army in the battle against Japan. In the meantime in Europe, Nazi Germany build the *V-1 flying bomb*[3], a cruise missile largely used in WWII. After the 1960s, it was Israel that gained interest in building unmanned vehicles for surveillance and reconnaissance purposes, of which the *Pioneer* is most known for its use in the Lebanon war in 1982. The US Army afterwards acquired the *Pioneer* for operations in the first Gulf War in 1990[4]. From there, it is a relatively small step to more contemporary surveillance and combat drones. US President George W. Bush and his successor Barack Obama gave the combat drone significant publicity by deploying the General Atomics *Predator* (first flight in June 1994[5]) and *Reaper* (first flight in February 2001[6]) in the "War on Terror" in Afghanistan and Iraq following the September 11 attacks in 2001. The main US battlefields where combat drones operate today are Somalia, Pakistan and Yemen. News about drone airstrikes in these regions dominate the actual debate from a rather singular military perspective.

But most people are at safe distance and therefore not directly confronted by the consequences of military drone operations. Nevertheless, their utility and actions are easy to imagine and reach us through captured imagery and videos, offering the spectator a first-hand view of the battlefield. The protocols of these remotely controlled operations add to this increasing degree of warfare abstraction among the public, and even the operators themselves. In a civilian context, drone applications have a rather recreational or commercial purpose, like photography and cinematography, event broadcasting, aerial mapping for agriculture and mining, wildlife conservation, and drone racing as the latest emerging hype. Inside the urban perimeter there are numerous attempts to introduce home delivery by drone, led by Amazon, who is actively trying to enter civil airspace since 2013 with its "Prime Air" project. Or the Swiss Post, that considers medical deliveries in remote regions affected by emergencies. Switzerland is one of the first countries that developed civilian drones with the OS4 project, a prototype quadcopter initiated by the Autonomous Systems Lab at EPFL Lausanne in 2004[7].

These endless applications and several utterly striking stories involving drones are highlighted in the journal. All selected articles derive from various international media outlets, in which ethical and political questions about drone usage are reviewed from different perspectives. The stories underline the importance of a better collective comprehension regarding the use of drones. And as long as restrictive legislation is still to be defined in detail, pilots will exploit the gaps in regulation, resulting sometimes in dangerous situations. Most notable examples are the numerous "almost collisions" — or "near misses" — that occur between drones and commercial airliners from 2013 onwards[8].

Then, the other part of *The Drone Chronicles* questions the exponential production of drones through an extensive catalogue. All models introduced to the market between 2001 and 2016 — either for commercial or military purposes — , are organised by release date, manufacturer and name, and characterised by their technical specifications. The manufacturers are listed by country in the catalogue's index, proposing a geopolitical comparative reading, equal to that of an atlas. The catalogue shows the production frenzy of an ever-growing industry, with market leaders like America, England, Israel and China competing with each other for the latest drone innovation.

In this rapidly evolving drone era, involved actors look at each other for guidance and seem to ignore their responsibility when a problem occurs. Federal regulators wait for government laws to be implemented, consumers continuously have advanced material at hand, but limited access to available fly-zones, and policy makers miss the urgency to propose laws that catch up with the industry's innovations. Events like the 'Drone and Public Safety Summit' in Brussels, the "European Drone Summit" in Stuttgart, or the EPFL "Drone Days" in Lausanne are welcoming platforms to start the discussion. Nevertheless, on this thin line of shifting responsibilities, a larger public debate is necessary to collectively decide what applications are of common interest, and to what degree we allow the buzzing drones access to our skies.

1 Jimmy Stamp, "Unmanned Drones Have Been Around Since World War I", https://www.smithsonianmag.com/arts-culture/unmanned-drones-have-been-around-since-world-war-i-16055939/

2 Preston Lerner, "The First Drone Strike – in 1944", in *Air & Space magazine*, https://www.airspacemag.com/military-aviation/drone-strike-180964753/

3 V-1 flying bomb, https://en.wikipedia.org/wiki/V-1_flying_bomb

4 Chris Cole, "Rise of the Reapers: A brief history of drones", https://dronewars.net/2014/10/06/rise-of-the-reapers-a-brief-history-of-drones/#_edn4

5 Ibid.,

6 General Atomics MQ-9 Reaper, https://en.wikipedia.org/wiki/General_Atomics_MQ-9_Reaper

7 http://www.homeofdrones.org/

8 List of UAV-related incidents, https://en.wikipedia.org/wiki/List_of_UAV-related_incidents

Involving drones in music; looking at *Kataster*
Brice Pauset

In the context of a huge cycle of works drawing a portrait of the technical, political and aesthetic genealogy that has led to our current historical situation, I wanted to tackle the question of surveillance and by extension that of the possibility of depicting or mapping the whole of reality.

Incidentally I always find it fruitful to brush up against subjects that are *a priori* incompatible with my profession as a composer, for it is through such well-nigh impossible instances of friction that genuinely new musical hypotheses can come into being.

I was initially very fascinated by the unique cinematographic set-up invented by Lars von Trier in *Dogville* (2003), a device aiming to superimpose the reality of a narrative situation on its own representation: even though the actresses and actors behave as if nothing is going on, the concrete places (the walls of a house for example) exist only partially in the form of abstract representations, thanks to white lines painted straight on to the black ground, itself thereby charged with abstraction. Thus the implacable logic of the story overrides the fate of each of the figures with a quite geometric precision, a precision reinforced still further by the way the set is devised.

The thing is that the question of place and our position in that place is no longer as simple as it used to be, to such an extent that the link between place and identity is broken. Two examples: "being somewhere", when we are physically there, involves all our affects, our impressions; however, a different way of "being somewhere" can also be summed up by polar coordinates (GPS coordinates). The other example is a bit more disturbing: When it comes to eliminating troublemakers the American administration is making ever more use of "clouds" of data within which the name of the target is often missing; instead there are telephone numbers, bank card numbers, polar coordinates, etc., intended to be fed into an algorithm which will in turn decide the suspect's fate. However, when it comes to moving on to eliminating the target, it is indeed a real flesh-and-blood person who disappears.

In the end it was after reading the philosopher Grégoire Chamayou's superb *Théorie du drone* that I was able to get the full measure of the important question of representation from the starting point of its most concise formulation: what does "being somewhere" mean today?

The now famous type MQ-1L *Predator* military drone, produced by the General Atomics company and used by the US Air Force from July 1995 to March 2018, transforms a great many aspects of conflicts, and indeed of life in society:

> the laws of war are changed in their entirety, since the symmetry of conflict is broken: it is the doctrine of a war with "zero casualties" — zero casualties for those who are on the side of the drones;

the gap between surveillance and punishment is very much reduced given that the same technical object oversees both matters;

the drone participates fully and completely in the ideology determined by the market economy since it reduces the costs of combat at three levels: political, technical and ethical.

The musical representation of such an abstract set of problems requires an appropriate musical approach, and first of all presupposes — in this adopting the main lesson of phenomenology — *not telling itself stories*... Thus representation, and hence duplication and misalignment between two instrumental groupings, one of them identifiable: a very large symphony orchestra, broadly adopting the instrumentation of Anton Bruckner's 9th symphony; the other in continuous transformation and change: five ensembles of instruments distributed right across the stage, amidst the orchestra, ensembles within which many windows open to self-representation are again opened up, like for example pianos used as resonators of the brass instruments, which themselves contaminate the brass section of the orchestra. But the ambiguity remains complete: is the orchestra represented by the ensemble, or is it the other way round? The whole piece is based on this permanent interplay of ambivalences, reciprocities, substitutions. Its title *Kataster* (German for "cadastre", or land register) refers explicitly to these problems.

It is for reasons of aesthetic affinity that I have chosen to transpose the Brucknerian orchestra into my work. However, an extra dimension in this choice was revealed a posteriori. We know that Bruckner regarded this symphony as a religious mystery directly linked to God. In its third movement (which partly serves as a formal matrix for my piece), tremendous moments really make us feel the whole weight of the image of the omnipotent *eye of God*. Might surveillance and combat drones be the individual optical cells of a monstrous insect's eye, simultaneously everywhere and nowhere, stealthy but concerned for our well-being and our safety, at all costs?

Running with your drone
Nicolas Nova

Among all the online communities that discuss drones, the one that fascinates me the most is certainly the runners using quadcopters. While I often come across drone users interested in aerial photography or racing drones, using a drone to do sport is a rather curious and unprecedented phenomenon for me. Why? First off, because such a use questions a practical point of view: how do you run with it? Do you control the device while thinking about your stride or your breathing? Does a coach operate the machine remotely? I also wonder about the main motivation to walk around with a sort of flying creature: is it a training device? A technological gadget designed to strut around in front of others? Some kind of security guard who offers a reassuring presence? These are all issues for which I have tried to seek answers in the online communities that are committed to this practice.

In the web journal "Florida Running & Triathlon", for example, there is this article entitled "Running With Drones. Your New Running Partner"[1] which describes the many advantages of jogging with a quadcopter. According to the author, the drone can set the pace for the race ("drone as pacekeeper") by flying in front of the riders at constant speed. This article also mentions the use of the drone as a stride companion ("drone as buddy"), like a friend or a dog, which gives both a reassuring presence and a motivation to surpass oneself ("drone as agent of peer pressure"). Finally, the quadcopter can be seen as a way of thinking about something other than yourself and distracting the runner from too much effort or possible suffering ("drone as welcome distraction").

However, the most interesting aspect is not really about the motivations for running with a flying vehicle. The most intriguing part of this usage of a quadcopter is its automatic dimension. You run, and the device flies either in front or just nearby you, without the need of a pilot.

Although this practice is certainly still uncommon, I do think it anticipates the advent of a new kind of personal assistant in the form of a flying drone. While this possibility is not new — think of the science-fiction show of the 1980s *Hard Time on Planet Earth* (1989) which featured an avatar hovering over the main protagonist — its current use in athletes' circles may be transposed to other domains. Recent trends in the field of surveillance or automation of daily life thus make plausible all kinds of intriguing services. Think about child-follow flying autonomous vehicles that could monitor and provides a real-time feedback for parents, dog-walker droids that would manage pet strolling and relief, hovering surveillance aides that may be to protect anxious pedestrians at night. Strange, but likely cases.

1 Running With Drones. Your New Running Partner? https://www.flrunning.com/blog/running-withdrones-your-new-running-partner-14237

A combination of a drone and a personal assistant such as *Alexa*, these flying companions will undoubtedly galvanize techno-optimists, renew the repertoire of absurd situations in everyday life, and, most certainly, reinforce the perplexity of contemporary Luddites.

2001 – 2016

Predator missile launch test totally successful

By SUE BAKER
Air Force News Service, 27 Februari 2001

WRIGHT-PATTERSON AIR FORCE BASE, OHIO – Aerospace history was made recently with the successful launch of a live missile from an unmanned aerial vehicle.

The Air Force's Predator Unmanned Aerial Vehicle (UAV) program is evolving from a nonlethal, reconnaissance asset, to an armed, highly accurate tank-killer, according to senior program officials from Air Combat Command (ACC) at Langley Air Force Base, Va., and Aeronautical Systems Center here.

"Capping a three-part series of demonstration flight tests on Feb. 21, Predator successfully aimed and launched a 'live' Hellfire-C, laserguided missile that struck an unmanned, stationary Army tank on the ground at Indian Springs Air Force Auxiliary Airfield near Nellis Air Force Base, Nev.," said Maj. Ray Pry, Predator program manager.

Flown by a pilot and sensor-operator from the 53rd Test and Evaluation Group at Nellis, who were located in a nearby Ground Control Station (GCS), Predator launched the missile using line-of-sight communication, inflicting heavy damage to the tank, Pry said.

The final flight, part of the Phase I feasibility demonstration that began in August, was preceded by two similar, completely successful Hellfire launches, Pry said.

"This first recorded missile launch from a UAV took place on Feb. 16," he said. "Equipped with a single, inert Hellfire-C missile, the Predator, using its line-of-sight communication band and infrared 'Kosovo' laser-ball, aimed and struck the tank-turret about 6 inches to the right of deadcenter, spinning the turret around about 30 degrees. It made a big, gray dent in the turret — just beautiful."

Following that first launch, the Predator/Hellfire launch team reviewed telemetry data and camera footage captured by the GCS crew and a helicopter from the Nellis Range, Pry said.

"We wanted to be sure that we had captured what we thought we had seen — that the stress and loads were within Predator's limits, and that the guides worked perfectly," he said. "With two shots planned for Feb. 21 using both satellite and LOS (Launcher Operation Station) communications links, we wanted to ensure we could use the satellite link to fire the missile."

With the initial weaponization feasibility tests successfully completed, Gen. John Jumper, ACC commander will review the results to determine when Phase II will begin, said Lt. Col. Tom Carlson, director of ACC's advanced weapons requirements branch.

"Phase II will take the Predator/Hellfire combination to more realistic, operational altitudes and conditions, including the challenge of a moving target," Carlson said. "This will complete the demonstration of the objectives we set down at the beginning of this process, to demo the technology, and prove its operational feasibility."

There are still some challenges ahead, the colonel said. "We need to do some re-engineering on the missile, to take it up to higher altitudes. Once we're given the 'green light' to proceed to Phase II — and all indications are that we will — it will require another symphony of players, brought together by Maj. Pry and his team, to execute the second round of demonstration flights."

"The bottom line is that we are taking a Hellfire missile, normally launched from an Army helicopter with its landing-skids 'in the trees,' or from the deck of a seaborne Navy carrier, flying under 2,000 feet, and asking it to fly at higher altitudes," Carlson said. "The recent Predator launches were done within the normal operating elevations for Hellfire."

Source: US Air Force, www.af.mil/news.

Taliban claim to have downed US unmanned spy plane

By JULIAN BORGER & RICHARD NORTON-TAYLOR
The Guardian, 24 September 2001

WASHINGTON – In what may prove to be the first shot of the new war in Afghanistan, the country's ruling Taliban militia claimed yesterday to have downed an unmanned spy plane, and the Pentagon admitted that one of its drones was missing.

The Taliban's official news agency, Bakhtar, said the aircraft was shot down over the Tashgurgan Pass in the northern Samangan province. The agency said Taliban officials were attempting to determine which country launched it.

In Washington yesterday, the defence secretary, Donald Rumsfeld, said the Pentagon had lost contact with a drone aircraft, but played down the Taliban's claims.

"The United States has, in fact, lost a … lost contact, I should say, with an unmanned aerial vehicle," Mr Rumsfeld told reporters after appearing on a Sunday talk show. "That happens from time to time in terms of the controls. We have no reason to believe it was shot down."

US forces have deployed electronic surveillance equipment in Uzbekistan, but it was not clear yesterday whether that equipment included unmanned spy planes like the Predator, which can fly at low altitude and send back detailed images of troop movements and landscape.

If the Taliban report is confirmed, the downed spy plane could have been mapping out entry routes into Afghanistan from Uzbekistan and checking Taliban defences.

Predators and other surveillance tools are expected to be used in the search for Osama bin Laden, prime suspect in the September 11 attacks.

The apparent loss of a spy plane comes as British military commanders put finishing touches today to an unprecedented war plan aimed at capturing or killing Bin Laden and his followers in Afghanistan.

General Sir Anthony Pigott, assistant chief of the defence staff for operations, who was in Washington last week, will brief ministers and senior military staff, including intelligence chiefs, on Britain's role in the planned assault.

Britain's joint force headquarters, which was to have been deployed to Oman for a huge exercise, Saif Sareea (Swift Sword), involving 24,000 troops and a naval taskforce of 28 ships, has been ordered to remain in Britain.

An extra 5,172 military US reservists were called up over the weekend, bringing to over 10,000 the total number recalled since September 11.

SAS troops are likely to be making contact with the Northern Alliance now fighting the Taliban.

An SAS squadron of about 50 troops is participating in the exercise in Oman and has been put on standby for action in Afghanistan.

The British naval taskforce includes a commando group of some 1,500 Royal Marines on board the helicopter carrier, HMS Ocean.

A similar number of troops from 16 Assault Brigade is due soon to leave Macedonia where they are on their way to completing a mission to collect weapons from ethnic Albanian guerrillas.

The SAS troops' knowledge of the movements and capability of Taliban forces, as well as minefields, could be crucial in any ground incursions in the north and north-west of Afghanistan, where, some reports say, Bin Laden has fled.

SAS and US special forces will also need local help for attacks on Bin Laden's bunkers and deep tunnels near Kandawest of Afghanistan, which can hide hundreds of people.

One of the most important roles of special forces, including the SAS, would be to provide intelligence from the ground via satellite communications, to guide pilots to specific targets. The planned assault will almost certainly include air strikes against known bases of Bin Laden.

Source: https://www.theguardian.com/world/2001/sep/24/afghanistan.terrorism18

Sources: US kills Cole suspect

CNN, 5 November 2002

SANAA, YEMEN – Six suspected al Qaeda members — including an al Qaeda chief wanted in the bombing of the USS Cole — were killed early Monday in Yemen when a CIA drone launched a "Hellfire" missile and struck the car they were traveling in, sources told CNN.

It was the first direct US strike against Osama bin Laden's al Qaeda network outside Afghanistan since the US-led war on terrorism began after the September 11, 2001, terrorist attacks.

Officials with the Central Intelligence Agency and Pentagon refused to discuss the report.

The sources who spoke to CNN said the Hellfire missile was launched from an unmanned Predator aerial vehicle. All six people in the car died, they said.

Video from the scene in Yemen's oil rich Marib province showed the car blown apart, with most of it reduced to black ash in the desert.

Sources identified one of the dead as Abu Ali, also known as Qaed Senyan al-Harthi, a former bin Laden security guard who was believed to have played a major role in the October 2000 attack on the destroyer Cole that killed 17 sailors.

Walid Al-Saqqaf, managing editor of the Yemen Times, told CNN that Ali was identified by a mark on his leg, which was blown off in the blast and found nearby.

He said Ali, who has been on the run and was believed to be harbored by tribesmen, has been the source of a massive hunt by security forces in Yemen. An attempt to capture him late last year failed. That botched attempt left more than a dozen security forces dead.

About 50 US Special Forces troops have been in the country training Yemeni security forces. There was no immediate indication they took part in the strike.

During a campaign rally in Arkansas, President Bush did not comment directly on the incident in Yemen but said the United States is pursuing "international killers."

"The only way to find them is to be patient and steadfast and hunt them down. And the United States of America is doing just that," Bush said.

Defense Secretary Donald Rumsfeld spoke of Ali at a Pentagon press briefing Monday when questioned about the attack.

"It would be a very good thing if he were out of business," Rumsfeld said.

Rumsfeld said the US-Yemen relationship "has been a good one and it's ongoing." He noted that Yemen's President Ali Abdullah Saleh visited the Pentagon and agreed to cooperate in the war on terrorism.

"As a result, we have some folks in that country that have been working with the government and helping them think through ways of doing things," Rumsfeld said.

"And it's been a good cooperation, and we've shared some information, and we think that over time it ought to be beneficial."

Rumsfeld said a number of al Qaeda members are known to be hiding in Yemen, slipping into the country by sea and through its sparsely populated border areas — what he said are used "advantageously by terrorists."

Source: http://edition.cnn.com/2002/WORLD/meast/11/04/yemen.blast/index.html

Iraqi drones not for WMD (weapons of mass destruction)

By LAUREN JOHNSTON
CBS News, 28 August 2003

BAGHDAD – Huddled over a fleet of abandoned Iraqi drones, US weapons experts in Baghdad came to one conclusion: Despite the Bush administration's public assertions, these unmanned aerial vehicles weren't designed to dispense biological or chemical weapons.

The evidence gathered this summer matched the dissenting views of Air Force intelligence analysts who argued in a national intelligence assessment of Iraq before the war that the remotely piloted planes were unarmed reconnaissance drones.

In building its case for war, senior Bush administration officials had said Iraq's drones were intended to deliver unconventional weapons. Secretary of State Colin Powell even raised the alarming prospect that the pilotless aircraft could sneak into the United States to carry out poisonous attacks on American cities.

The administration based its view on a Central Intelligence Agency finding that Iraq had renewed development of sophisticated unmanned aerial vehicles — UAVs — capable of such attacks. The Pentagon's Defense Intelligence Agency also supported this conclusion. While the hunt for suspected weapons of mass destruction — and the means to deliver them — continues, intelligence and defense officials said the CIA and DIA stand by their prewar assertions about Iraqi drone capabilities, some of which Powell highlighted in his Feb. 5 presentation to the U.N. Security Council.

But the Air Force, which controls most of the American military's UAV fleet, didn't agree with that assessment from the beginning. And analysts at the Pentagon's Missile Defense Agency said the Air Force view was widely accepted within their ranks as well.

Instead, these analysts believed the drones posed no threat to Iraq's neighbors or the United States, officials in Washington and scientists involved in the weapons hunt in Iraq told The Associated Press.

The official Air Force intelligence dissent is noted in the October 2002 National Intelligence Estimate on Iraq's weapons programs, parts of which were declassified last month as the Bush administration tried to defend its case for war.

"We didn't see there was a very large chance they (UAVs) would be used to attack the continental United States," Bob Boyd, director of the Air Force Intelligence Analysis Agency, said in an AP interview. "We didn't see them as a big threat to the homeland."

Boyd also said there was little evidence to associate Iraq's UAVs with the country's suspected biological weapons program. Facilities weren't in the same location and the programs didn't use the same people.

Instead, the Air Force believed Iraq's UAV programs were for reconnaissance, as are most American UAVs. Intelligence on the drones suggested they were not large enough to carry much more than a camera and a video recorder, Boyd said.

Postwar evidence uncovered in July in Iraq supports those assessments, according to two US government scientists assigned to the weapons hunt.

"We just looked at the UAVs and said, 'There's nothing here. There's no room to put anything in here,'" one of the scientists said.

The wingspan on drones that Iraqis showed journalists in March measured 24.5 feet and the aircraft were built like large, white model airplanes.

The US scientists, weapons experts who spoke on condition of anonymity, reached their conclusions after studying the small aircraft and interviewing Iraqi missile experts, system designers and Gen. Ibrahim Hussein Ismail, the Iraqi head of the military facility where the UAVs were designed. None of the Iraqis questioned are in US custody.

While the weapons hunters can't be sure they've recovered all of Iraq's UAVs, the evidence amassed so far, coupled with the interviews, has led them to believe that none of the drones are designed for unconventional weapons. Iraqis involved in the program have insisted the drones were for reconnaissance and electronic jamming.

Some UAVs were kept north of Baghdad. Weapons hunters found some drones in better shape than others with the most important finds located at a facility in the capital, the US scientists said. Weapons hunters hauled them back to their base on the outskirts of the Baghdad International Airport where the parts were analyzed.

The unproven US assertion regarding Iraq's UAV programs is one among many.

American weapons hunters, like their U.N. counterparts, haven't reported finding any chemical, biological weapons or nuclear weapons in Iraq so far.

The lack of success in uncovering unconventional weapons, after warnings that Iraq posed an immediate danger, has led critics and some former government analysts to suggest the administration exaggerated the threat posed by Saddam.

Boyd said the Air Force's dissent was handled fairly, and that his analysts did not feel pressured to alter their position. "Our view was fully aired in the process," he said.

The Bush administration has made public some of what led it to believe the UAVs were for biological or chemical weapons attacks.

Before the war, US intelligence agencies learned that officials with Iraq's UAV program tried to buy commercially available route-planning software that was packaged with electronic maps of the United States, according to the declassified portion of the National Intelligence Estimate.

This discovery was interpreted by some analysts as a sign Iraq was trying to plan UAV bombing runs over the United States. But Boyd said Air Force analysts were unconvinced because maps are frequently bundled with such software.

At the United Nations in February, Powell told the world Iraq had test-flown a UAV well beyond a 93-mile limit allowed under U.N. rules. But both reconnaissance and offensive aircraft would need to travel long distances, Boyd said.

Compared to other agencies, Boyd said the Air Force relied more on information from reconnaissance satellites and less on defectors, in accessing Iraq's UAVs.

Saddam's regime had experimented with remotely controlled jet aircraft modified for biological and chemical attacks before the 1991 Persian Gulf War, but U.N. inspectors found no evidence that program had been successful.

Boyd said attempts in the mid-1990s by Saddam's regime to convert an L-29 jet trainer into a dispersal system were abandoned. US weapons hunters also studied jet trainers found in northern Iraq but found no evidence they had been converted into biological or chemical weapons carriers, they said.

Source: https://www.cbsnews.com/news/iraqi-drones-not-for-wmd/

Drone school, a ground's-eye view

By NOAH SHACHTMAN
Wired, 27 May 2005

Fort Huachuca, Arizona – The loping, wooden barracks here look pretty much the same as they did in the 1880s, when soldiers marched off to chase Geronimo and his Apache guerillas. From the open, double-story porches, you can still see the adobe home where commanders planned one of the US Army's final cavalry charges on horseback, 30 years after that.

Just five miles away, on a lonely airstrip on the base's northwestern edge, 225 soldiers, national guardsmen and reservists are training for the most modern kind of warfare. They're learning to fly the unmanned aerial vehicles, or UAVs, which have become so critical to the battle for Iraq. By the end of the year, most of these freshly minted pilots will be in hot zones like Baghdad and Fallujah, using their robot planes to spy on insurgents and keep watch over American troops below.

In March, Wired magazine sent me to Fort Huachuca, to learn more about this new breed of pilot.

Most of the teaching here at Fort Huachuca, from federal flight regulations to the operation of the drones' infrared cameras, is done by outside contractors. Bill Hempel's specialty is training "external pilots" — the guys who use twin joysticks to take off and land the 23-foot-long, split-tailed Hunter UAVs. It's kind of an unmanned master class; only soldiers who already know how to keep a drone in the air — which is just a matter of pointing and clicking, really — can sign up for external pilot duties.

"We take these Army guys who don't know the front end of an airplane from the back, and we teach them from scratch all the aviation they need to make them pilots," he explains.

Hempel has been flying remote-controlled airplanes for 35 years. Five times, he's won the US national championship in aeronautical acrobatics. All of which makes Hempel absolutely indispensable in an Army where soldiers have to pilot drones from the ground. He's been a UAV instructor at Fort Huachuca since 1993.

Hempel starts his students out on a PC simulator, to get them thinking in three dimensions. Then, they move on to real-life radio-controlled planes and one-third-scale Hunters, for 40 hours of flight time apiece. After another 40 hours taking a full-sized Hunter up and down, the soldiers are certified external pilots.

By then, they're usually hooked on model airplanes, too. "I want to get into the (aeronautical acrobatics) competitions, too," says Spc. Daryl Orcelletto, who recently returned from a year of flying drones over Iraq. "I tell Mr. Hempel, 'I want to learn everything you know.'"

But even with all of Hempel's training, once his former students are in the field, they give up the joysticks when they're asked to make a tricky landing. Instead, they get contractors to do the job. Jimmy Vargas spent four years in the Balkans and one year in Iraq with the Army's 15th Military Intelligence Battalion, as part of a five-man Northrop Grumman logistics support team.

Officially, the former sergeant and Hunter pilot was there to "provide the institutional memory" to a revolving array of commanders, he says — to tell the guys in uniform what the machine was really capable of. But when the weather turned ugly, "certain soldiers were not as comfortable flying," Vargas recalls. "They'd ask me, 'How do you feel? Can you do this launch?'"

More than once a week in Iraq, he would. "In my unit," he says, "the only difference between us (the contractors) and them (the people in uniform) was that we didn't have weapons."

UAV USO

When drone devotees talk about UAVs' importance to the American war machine, they usually talk about the pilots' lives they save, or about the tactical advantages in seeing a battle unfold from above. They rarely get around to mentioning the drones' entertainment value.

Digital footage from the robot planes is now routinely sent everywhere the military's network extends, which means soldiers far removed from the front lines finally get to see a little action in real time. "It's like a video game," says one analyst who served at US Central Command headquarters in Camp As Sayliyah in Qatar. "It can get a little bloodthirsty. But it's fucking cool."

July 22, 2003, was a particularly gory — and particularly eye-popping — day, Sgt. Frederick Lewis remembers. The Fort Huachuca instructor won a Bronze Star in Iraq for his Hunter UAV external piloting. (Northrop contractor Jimmy Vargas served with his unit, and the two now share an apartment in Arizona.)

From an airfield about five miles away, Lewis watched as a Hunter peered down on a sprawling villa on the northern outskirts of Mosul. Troops from the 101st Airborne Division were beginning to mass outside of it. So were Bradley Fighting Vehicles and Humvees — dozens of them. "That's when we knew something big was about to go down," he says.

Forty soldiers piled into the 15-by-12-foot tent serving as the local command center, squirming for a view of the lone, 25-inch flat-panel screen. Then, the 101st began raining hell on the villa, with cannons and grenades and missiles fired almost at point-blank range. In the tent, they whooped with every explosion. "It was like a Super Bowl party in there," Lewis says with a barely contained grin. The cheers welled up again a few hours later, when the troops found out who was inside the house: Uday and Qusay Hussein, Saddam's genocidal sons.

Borderline convert

When his boss first brought up the subject 18 months ago, US Border Patrol agent Floyd Robbins didn't want to know about any robot airplanes. Robbins already had enough on his plate, running a Border Patrol mapping team here in the country's busiest corridor for illegal immigration. Then he saw the drones in action above the mountain passes and agave-covered gulches that separate Arizona from Mexico. Now, the 11-year veteran gets giddy like a rookie, showing off what his unmanned craft can do.

We're in Robbins' office — a dingy, haphazard collection of desks attached to a Fort Huachuca helicopter hangar. He smiles as he cues up footage taken in mid-January from a Hunter drone. In it, about 80 immigrants are huddled together on a ridge line in the Huachuca Mountains. From the lower-right and upper-right corners of the screen, a half-dozen agents begin to close in.

Everyone looks like germs, like ants, from the Hunter's 15,000-foot point of view. Especially when the ant hill breaks apart, and everybody scatters in a dozen different directions.

Before they got drones, agents were pretty much helpless when this happened. "You'd get what your hands could hold, heh heh, heh heh," Robbins laughs. "Maybe like two or three people."

But with the Hunter overhead, the migrants never leave the agents' sights. Robbins spots one group taking cover underneath a bush. He radios the guards, who swoop in and bind the illegals up. Meanwhile, a set of more than 20 aliens scampers into a canyon.

"We couldn't get to them right then. So we tracked them with the UAV, came back the next day, and got 'em, hee heh heh," Robbins explains, chuckling a bit. "With the UAV, there's nowhere for them to hide. Once it locks on, they're gonna get busted."

The drone was just a loaner — part of a six-month experiment to see if UAVs could help the Border Patrol. During that stretch, Robbins and a team of eight agents were able to catch more than 1,200 illegals and seize 2,700 pounds of marijuana, thanks to the drones. Now, there's legislation in Congress to buy the Border Patrol some drones outright. Robbins can't wait.

Source: https://www.wired.com/2005/05/drone-school-a-grounds-eye-view/

High hopes for drone in LA skies

By PETER BOWES
BBC News, 15 January 2006

Los Angeles – Unmanned aerial vehicles (UAVs), commonly known as drones, are about to be launched for the first time by the police in Los Angeles.

UAVs have long been used by the military in war zones such as Iraq or Afghanistan. But the technology has been adapted for domestic use and could revolutionise the way law enforcement agencies carry out surveillance and rescue operations.

The Los Angeles Sheriff's Department (LASD) has been experimenting with a drone called SkySeer, which it intends to put into service later this month.

SkySeer looks like a remotely controlled model aircraft. It has a wingspan of 6.5 feet (1.98m) and weighs 4 pounds (1.81kg). A camera is attached to its belly and a small battery powers the drone.

Almost invisible

"It has a video link that sends data in real time down to our ground station — the operator can then see, in real time, what it's seeing," explains SkySeer inventor Sam De La Torre, from Octatron Inc — a surveillance technology firm.

The SkySeer has been designed for quick and easy use by police officers on the street. It can be folded up and stored like a tent in a backpack.

"Within five minutes he can have the aircraft assembled,' says Mr De La Torre"

"You just push the take-off button, the motor starts up and you throw it."

The UAV can fly at any height. At 250 feet above the ground, it can clear a 25-storey building and is almost invisible.

The Sheriff's Department is keen to start using the drone in situations where conventional crime-fighting is either impractical or too expensive.

At a cost of approximately $25,000-$30,000 (£13,400-£16,000), the UAV is considerably cheaper than a helicopter. But the device's practical applications are generating the most excitement amongst officers.

Pinpointing victims

"It provides several things that we can't get other ways," says Commander Charles Heal, head of the LASD's technology exploration project.

The UAV's ability to hover in virtual silence over an accident or crime scene, without any risk to a pilot, provides both a tactical and economic advantage.

It is envisaged that SkySeer will be put to use when children go missing down a hillside in difficult terrain.

To save time and minimise the risk to rescuers, the UAV will be used overhead to pinpoint the location of a victim.

"It has different cameras — colour, low light and even infra-red — and so as a result of that we can even find heat signatures that are coming through the bushes and overhead," says Commander Heal.

With burglaries, the police say the SkySeer will be used get an aerial view of a building where someone is believed to have broken in through the roof.

The conventional approach is to call the fire department to bring in ladder trucks, allowing officers physically to climb onto the top of a building.

"If the suspect really wants to hurt you, your head is the first thing that he sees. Now we'll have the ability to actually to fly this over and see if it is even worth doing a containment."

The UAV utilises an onboard compass and GPS system for its command and control. It flies to a location that is predetermined by the operator on a laptop.

The developers are working on a so-called cyber command post to enable images to be viewed, anywhere in the word, in real time.

"If we're flying over hazardous material or something that we can't recognise, we can have a subject matter expert, maybe not even in the country, in a different time zone, that is actually watching the exact same information that we're getting."

'Big brother' surveillance?

"We will be able to incorporate his subject matter expertise into our decision making process," says Commander Heal.

The SkySeer will also be used to back up officers on the ground if they are pursuing a suspect on foot. Flying at a speed of about 30 mp/h (48 km/h), the police believe it will be impossible for a suspect to outmanoeuvre the UAV.

"You simply point the camera at him and keep following."

The Los Angeles Sheriff's Department has only one prototype SkySeer at the moment.

When it goes into service, the force's SWAT [Special Weapons and Tactics] unit will carry out the initial evaluation in real-life situations.

Commander Heal is quick to point out that it is not their intention to launch 'big brother' style surveillance operations.

"There's no place in an urban environment that you can go to right now that you're not being looked at with a video camera and you have nothing to fear from your own government — you are being watched by your fellow citizens," he says.

"The only time that this is ever going to be operational is in some kind of emergency condition."

Source: http://news.bbc.co.uk/2/hi/americas/5051142.stm

Eye in the sky: police use drone to spy on V festival

By JAMES RANDERSON
The Guardian, 21 August 2007

WESTON-UNDER-LIZARD, UK – Police used a remote-controlled spy drone to watch crowds at the V festival at the weekend, the first time the technology has been used at a major public event.

The 70cm-wide flying surveillance device, fitted with high-resolution still and colour video cameras as well as infrared night vision capability, was used to keep tabs on people thought to be acting suspiciously in car parks and to gather intelligence on individuals in the crowd.

Staffordshire police said the drone's images did not lead directly to any arrests, but one reason for using it had been to deter would-be thieves. It was not flown over the main arena because of fears that a crash might cause injuries.

The battery-operated drone's four carbon-fibre rotors are so quiet they cannot be heard from the ground once it is higher than 50 metres, and at 100 metres up it cannot be seen with the naked eye. It can fly 500 metres high, but the Civil Aviation Authority has set an operating limit of 120 metres. The vehicle, which takes off vertically, can be flown even when out of sight, because it beams images back to video goggles worn by the operator.

Since May, Merseyside police have been using two drones — originally developed by a German company for military use — to police public order situations and prevent antisocial behaviour. One tactic is to fly the drone over groups of young people causing a nuisance in parks. The force has also used it for covert surveillance.

The West Midlands fire service plans to use drones to get a bird's eye view of buildings on fire. "Being able to look down on the scene will allow us to get a full picture of the incident and the surrounding environment, which will aid incident commanders to make vital, potentially life-saving decisions," said the deputy chief fire officer, Vij Randeniya.

The Metropolitan police is interested in using drones to police the Olympics. Firearms officers also believe they could be useful in a standoff with armed criminals

MW Power, the company that distributes the technology in the UK, plans to improve the drone's capability by adding a so-called "smart water" spray — a liquid infused with unique artificial DNA sequences which can be squirted on to a suspect from above. It infuses their clothes and skin and the DNA code can be used later to identify them.

There is no legal barrier preventing a private security firm or a paparazzo photographer from using the technology, but MW Power said that it was only licensing the vehicle to customers from the military or emergency services. It costs less than £1,000 a month to lease — an amount that would buy less than an hour's use of a conventional helicopter.

Some experts fear it represents an unwarranted intrusion of privacy. "We should find out whether the public wants this," said Noel Sharkey, an expert in robotics at Sheffield University who is worried about the increasing use of robotic vehicles in military conflicts and policing. While most people would support its use to catch car thieves, the technology could be put to more draconian uses in future, he said.

"How long will it be before someone gets Tasered from the air for dropping litter, or even for relieving themselves down an alleyway under cover of night?"

Source: https://www.theguardian.com/uk/2007/aug/21/ukcrime.musicnews

Killer drone's Iraq debut

By DAVID AXE
Wired, 25 July 2008

BAGHDAD – The MQ-9 Reaper, a bigger version of the famed Predator drone, flew its first mission over Iraq on July 18, according to the Air Force. The armed Reaper deployed to Afghanistan in 2007 and first saw combat in the fall of that year. British officers compared the 5-ton bird to a "mini A-10." The Brits are buying around a dozen Reapers of their own for Afghanistan service.

Reaper's Iraq deployment is part of Secretary of Defense Robert Gates' drive to get more surveillance systems into the fight. The Pentagon's goal was to have enough deployed Predators and Reapers by 2010 to maintain 21 round-the-clock "orbits," each requiring three or four drones. By raiding the training unit at Creech Air Force Base, Nevada, the Air Force beat the deadline by two years. In May, the service announced that it had 24 drone orbits for the "war on terror," including Iraq, Afghanistan, Pakistan and (presumably) Somalia.

But taking 'bots from Creech has put kinks in the training system and possibly contributed to several recent crashes.

Defense firm Raytheon says better control stations are the solution. "The [current] Predator ground station displays are like an engineering diagnostics station, with complicated menus and 'M-keys' with functions that are easily confused," said Katie Heilner, a Raytheon engineer. The company is offering new, simplified "Universal Control Stations" that it says will reduce the robot accident rate.

Regardless, Air Force is now aiming to deploy as many as 50 orbits by the end of this year. Lt. Gen. Mike Peterson said he wants most of them to be Reapers, but right now the Air Force has only a dozen of the bigger birds.

Source: https://www.wired.com/2008/07/killer-drones-i/

Mysteries surround Afghanistan's stealth drone

By DAVID HAMBLING
Wired, 12 April 2009

KANDAHAR, AFGHANISTAN – Earlier this year, blurry pictures were released by the French magazine *Air & Cosmos* of a previously unknown stealth drone taken at Kandahar in Afghanistan. The photos, snapped in 2007, prompted a wave of speculation about the classified aircraft. That speculation grew even more intense this week, when a blog belonging to the French newspaper *Libération* released an even better photograph. But while the new picture may answers some questions, it also creates a heap of new mysteries. Chief among them: Why use such a fancy, stealthy aircraft in Afghanistan? The Taliban have neither the radar to spot the plane, nor the weaponry to shoot it down.

The lines of the drone clearly indicate a stealth design slightly reminiscent of the B-2A Spirit bomber, but smaller. Over on Ares, veteran aviation expert Bill Sweetman describes the wingspan as being perhaps eighty feet, and notes "One important detail: the overwing fairings are not B-2-like inlets, but cover some kind of equipment - satcoms on one side, perhaps, and a sensor on the other."

The aircraft, which Bill has dubbed the Beast of Kandahar, is widely believed to be a product of Lockheed's celebrated Skunk Works, home of the F-117 Nighthawk stealth fighter. The Beast bears some resemblance to the Skunk Works' Polecat drone, revealed in 2006. This was a private venture, costing some $27 million of Lockheed's own money, designed to operate stealthily at high altitude and at supersonic speed. However, the Beast is not Polecat, as one look at the exhaust will indicate.

The Beast has also been identified with the covert Desert Prowler program, identified by black ops spotter Trevor Paglen. The Desert Prowler's patches include the phrases "alone and unafraid" and "alone and on the prowl" as well as the figure of a wraith taken from an album cover by Insane Clown Posse. The wraith is said to represent the Grim Reaper...peculiar as it may seem, Paglen has shown that a remarkable amount of information can be gleaned from Black Ops patches and has written a book on the subject.

Meanwhile, there is considerable speculation on the Secret Projects Forum about technical aspects of the Beast, including attempts to calculate its exact size. The landing gear may have been borrowed from another aircraft, a common practice in drone circles to keep costs down.
If the gear could be identified, it might say a lot about the size and weight of the Beast.

However, having established that there really is a stealthy US drone operating out of Kandahar, the big question is what is it doing there? The Taliban do not have radar, so why deploy an expensive, stealthy drone when conventional models like the Predator and Reaper work so well? And what's the point of having a high-level, strategic craft in that theater?

There has of course been plenty of speculation. Much of it is focused on the idea that while it is based in Kandahar, the Beast may be carrying out missions outside of Afghanistan, with Iran and Pakistan both being possible candidates. For both of those radar stealth could be an important asset, and the beast may be carrying out signals-intercept or other tasks (looking for traces of nuclear material?).

This might make more sense than local operations: why risk an expensive, scarce black drone and run the chance of it being uncovered if you can do the job with other aircraft. On the other hand, if the Beast is a top-secret craft on a top-secret mission, why leave it out where it can be photographed?

One note of caution: the provenance of the photograph is not known, and it was published "without guarantee of origin." So it might be some deliberate disinformation to put black plane-spotters off the scent of the real Beast.

Source: https://www.wired.com/2009/12/mysteries-surround-afghanistans-stealth-drone/

US now trains more drone operators than pilots

By EDWARD HELMORE
The Guardian, 23 August 2009

ARLINGTON COUNTY, VIRGINIA – As part of an expanding programme of battlefield automation, the American air force has said it is now training more drone operators than fighter and bomber pilots.

In a controversial shift in military thinking — one encouraged by the confirmed death of Pakistani Taliban leader Baitullah Mehsud in a drone-strike on 5 August — the air force is looking to hugely expand its fleet of unmanned aircraft by 2047.

Three years ago, the service was able to fly just 12 drones at a time; now it can fly more than 50. At a trade conference outside Washington last week, military contractors presented a future vision in which pilotless drones serve as fighters, bombers and transports, even automatic mini-drones which attack in swarms.

Five thousand robotic vehicles and drones are deployed in Iraq and Afghanistan. By 2015, the Pentagon's $230bn (£140bn) arms procurement programme Future Combat Systems expects 15% of America's armed forces to be robotic. A recent study 'The Unmanned Aircraft System Flight Plan 2020-2047' predicted a boom in drone funding to $55bn by 2020 with the greatest changes coming in the 2040s.

"The capability provided by the unmanned aircraft is game-changing," said General Norton Schwartz, the air force chief of staff. "We can have eyes 24/7 on our adversaries."

Some analysts view the Flight Plan study a virtual death knell for the pilot profession and predict the F-22s successor, the F-35 Joint Strike Fighter, could be the last piloted fighter programme.

Colonel Eric Mathewson, who directs the air aorce task force on pilotless aerial systems, has sought to downplay the studies' most futuristic predictions. "We do not envision replacing all air force aircraft with UAS (unmanned aircraft systems)," he said.Currently airborne drones are directed by trained pilots who then return to their assigned aircraft. This year, the service started training career drone operators with no airborne experience. They operate out of cubicles with eight video screens.

"It is safe to say most pilots will always miss getting back in the air," said Lieutenant Colonel Daniel Turner, who trains pilots. "But we see where the Air Force is going."

In *Wired for War*, author Pete Singer speculates the machines are harbingers of a new era of "cost-free war".

"It's an historic change," said Singer. "Going to war has meant the same thing for 5,000 years. Now going to war means sitting in front of a computer screen for 12 hours. Then you go home and talk to your kids about their homework."

Source: https://www.theguardian.com/world/2009/aug/23/drones-air-force-robot-planes

Drone makes first UK 'arrest' as police catch car thief hiding under bushes

By LIZ HULL
The Daily Mail, 12 February 2010

Litherland, UK – It has been nicknamed the flying saucepan and looks an unlikely weapon in the war against crime. But yesterday it emerged that a suspected car thief had become the first person to be arrested in Britain thanks to the help of this miniature remote-controlled helicopter.

The Air Robot or drone was deployed by Merseyside police after officers lost the alleged offender who had escaped on foot in thick fog.

Using the device's on-board camera and thermal-imaging technology, the operator was able to pick up the suspect through his body heat and direct foot patrols to his location.

It led officers to a 16-year-old youth, who was hiding in bushes alongside the Leeds-Liverpool canal, in Litherland.

The drone, which measures 3ft between the tips of its four carbon fibre rotor blades, uses unmanned aerial vehicle (UAV) technology originally designed for military reconnaissance.

The battery-powered device can have a range of cameras attached to its main body, including CCTV surveillance or thermal imaging cameras.

It is designed to hover almost silently above crime scenes and send live footage to officers on the ground, but the unit can also 'perch and stare' from a solid platform, allowing the operator to capture hours of footage from a hidden vantage point.

> *Source: https://www.dailymail.co.uk/news/article-1250177/Police-make-arrest-using-unmanned-drone.html*

President Obama's drone joke misses mark

By PATRICK GAVIN
Politico.com, 4 May 2010

WASHINGTON – It's never too late to over-analyze the stand-up routines delivered at Saturday's White House Correspondents' Dinner by both President Barack Obama and Jay Leno. So here's another bit of thinking: Obama's joke regarding the Jonas Brothers wasn't funny at all.

First, the joke as originally told: "The Jonas Brothers are here. (Applause.) They're out there somewhere. Sasha and Malia are huge fans. But, boys, don't get any ideas. (Laughter.) I have two words for you — Predator drones. (Laughter.) You will never see it coming. (Laughter.) You think I'm joking. (Laughter.)"

Why the criticism from some corners? Well, Predator drones are a somewhat controversial part of America's counter-terrorism program, having resulted in the deaths of innocent civilians in Pakistan (no one knows how many for sure — the CIA's drone program is secret).

"It's funny because Predator drone strikes in Pakistan have killed literally hundreds of completely innocent civilians, and now the president is evincing a casual disregard for those lives he is responsible for ending by making a light-hearted joke about killing famous young celebrities for the crime of attempting to sleep with his young daughters," said Salon's Alex Pareene

The American Prospect's Adam Serwer said that the joke undermined much of Obama's good work throughout the Muslim world. "So you have to wonder why in the world the president's speechwriters would think it was a good idea to throw a joke about Predator drones into the president's speech during the White House Correspondents' Dinner, given that an estimated one-third of drone casualties, or between 289 and 378, have been civilians. It evinces a callous disregard for human life that is really inappropriate for a world leader, especially a president who is waging war against an enemy that deliberately targets civilians. It also helps undermine that outreach by making it look insincere."

Conservative blogger William Teach wrote, "Again, just imagine Bush telling THAT joke and what the media would be printing the following day."

"Let's be honest, fellow progressives," said the Philadelphia Daily News' Will Bunch, "we'd be all over Bush if he made the same 'Predator drone' joke Obama told last night."
Teach and Bunch are onto something: When President George W. Bush cracked a joke about missing weapons of mass destruction during his 2004 White House Correspondents' Dinner remarks, the 43rd president got heat for making light of a very serious topic. Then-DNC chair Terry McAuliffe called Bush's joke "inappropriate to the thousands of people obviously who have been wounded over there."

Source: https://www.politico.com/click/stories/1005/obamas_drone_joke_misses_mark.html

Taliban says US drone attacks 'temporarily' hindering insurgency

By MARK TRAN
The Guardian, 23 July 2010

LONDON – The Taliban have admitted that US drone attacks have disrupted their operations in Afghanistan but insist it will be only for the short term.

Under Barack Obama, the US has stepped up the use of unmanned aircraft to target Taliban leaders and foreign fighters sheltering in neighbouring Pakistan, particularly in the border region of North Waziristan and South Waziristan.
"In the short term, yes, you can say it has caused us some difficulties because of the martyrdoms and realignment of our ranks," a Taliban spokesman, Muhammed Umer, told the BBC. "But our command and control system is very strong and well established, so we won't be affected for long," he said, adding that anger at drone attacks was drumming up new recruits. "Instead we get new courage, becoming more powerful with the flow of new blood."

Analysts say the increase in drone attacks has forced a change in the pattern of militant behaviour in the tribal area, with Taliban and foreign fighters more careful about gathering in large groups and tending to move on from locations more quickly.

In the most recent attack, Pakistani security officials last week said that between 10 and 14 "militants" were killed in a Predator strike, although no senior Taliban or al-Qaida leaders were reported killed. It was the first strike reported inside Pakistan this month, coming after a 16-day hiatus — the longest delay between strikes recorded since the US stepped up its the air campaign at the end of July 2008.

So far this year, the US has carried out 46 strikes in Pakistan, with all but three in North Waziristan. The other two strikes took place in South Waziristan and the tribal agency of Khyber. The US is well on its way to exceeding last year's strike total in Pakistan. In 2009, the US carried out 53 strikes in Pakistan; and in 2008, the US carried out 36.

As for casualties, 700 people reportedly have been killed in such attacks under Obama, compared with slightly fewer than 200 from under his predecessor, George Bush.

> *Source: https://www.theguardian.com/world/2010/jul/23/taliban-drone-attacks-pakistan*

US building secret drone bases in Africa, Arabian Peninsula, officials say

By Craig Whitlock and Greg Miller
The Washington Post, 20 September 2011

WASHINGTON – The Obama administration is assembling a constellation of secret drone bases for counterterrorism operations in the Horn of Africa and the Arabian Peninsula as part of a newly aggressive campaign to attack al-Qaeda affiliates in Somalia and Yemen, US officials said.

One of the installations is being established in Ethiopia, a US ally in the fight against al-Shabab, the Somali militant group that controls much of that country. Another base is in the Seychelles, an archipelago in the Indian Ocean, where a small fleet of "hunter-killer" drones resumed operations this month after an experimental mission demonstrated that the unmanned aircraft could effectively patrol Somalia from there.
The US military also has flown drones over Somalia and Yemen from bases in Djibouti, a tiny African nation at the junction of the Red Sea and the Gulf of Aden. In addition, the CIA is building a secret airstrip in the Arabian Peninsula so it can deploy armed drones over Yemen.

The rapid expansion of the undeclared drone wars is a reflection of the growing alarm with which US officials view the activities of al-Qaeda affiliates in Yemen and Somalia, even as al-Qaeda's core leadership in Pakistan has been weakened by US counterterrorism operations.

The US government is known to have used drones to carry out lethal attacks in at least six countries: Afghanistan, Iraq, Libya, Pakistan, Somalia and Yemen. The negotiations that preceded the establishment of the base in the Republic of Seychelles illustrate the efforts the United States is making to broaden the range of its drone weapons.

The island nation of 85,000 people has hosted a small fleet of MQ-9 Reaper drones operated by the US Navy and Air Force since September 2009. US and Seychellois officials have previously acknowledged the drones' presence but have said that their primary mission was to track pirates in regional waters. But classified US diplomatic cables show that the unmanned aircraft have also conducted counterterrorism missions over Somalia, about 800 miles to the northwest.

The cables, obtained by the anti-secrecy group WikiLeaks, reveal that US officials asked leaders in the Seychelles to keep the counterterrorism missions secret. The Reapers are described by the military as "hunter-killer" drones because they can be equipped with Hellfire missiles and satellite-guided bombs.

To allay concerns among islanders, US officials said they had no plans to arm the Reapers when the mission was announced two years ago. The cables show, however, that US officials were thinking about weaponizing the drones.

During a meeting with Seychelles President James Michel on Sept. 18, 2009, American diplomats said the US government "would seek discrete [sic], specific discussions ... to gain approval" to arm the Reapers "should the desire to do so ever arise," according to a cable summarizing the meeting. Michel concurred, but asked US officials to approach him exclusively for permission "and not anyone else" in his government, the cable reported.

Source: https://www.washingtonpost.com/world/national-security/us-building-secret-drone-bases-in-africa-arabian-peninsula-officials-say/2011/09/20/gIQAJ8rOjK_story.html?utm_term=.63d09a78d4db

Using a phone to fly a drone

By DAVID L. CHANDLER
MIT News Office, 8 November 2011

Cambridge, Massachusetts – Imagine controlling an airplane in flight just by holding your iPhone out in front of you: tilting it in the direction you want the plane to travel, or raising it to make the plane fly higher. Or tapping a point on a map on the screen, and having the plane automatically fly to the designated spot.

Now, imagine if the plane itself were a continent away from where you're doing this iPhone-based controlling.

This is not, in fact, some science-fiction vision of the future — it actually happened this summer, with people at Boeing's Seattle research and development center controlling a small rotorcraft, or Unmanned Aerial Vehicle (UAV), as it flew around an athletic field on the MIT campus in Cambridge — some 2,500 miles away.

But the most significant thing about this long-distance control, explains MIT associate professor of aeronautics and astronautics Mary "Missy" Cummings — who designed the controller along with her students — is how easy it is for even an inexperienced person to fly the plane. The control system designed by Cummings and her students is so simple and intuitive that operators can take charge of flying the plane after just a few minutes of instruction. By comparison, soldiers who control existing UAVs must undergo a comprehensive, months-long training program.

Cummings, who directs the Humans and Automation Lab at MIT, focuses her research on how to make control systems that are easy for people to learn and use. In principle, she says, the control system she and her team have created for smartphones could be used to control any aircraft, even a jumbo jet. In practice, it could easily replace the control systems not only for military drones, but for UAVs used by emergency personnel: for example, to track the progress of a forest fire in a remote area from a safe distance.

Source: http://news.mit.edu/2011/iphone-drone-control-1108

Iran shows off captured US drone

By DAVID BLAIR & ALEX SPILLIUS
The Telegraph, 8 December 2011

TEHRAN – Iran triumphantly displayed one of America's most advanced intelligence-gathering aircraft on Thursday after a spy drone crashed on its territory, leading Russia and China to ask to inspect its technology.

The RQ-170 Sentinel, supposedly the CIA's unseen "eye in the sky", capable of beaming back a trove of imagery and electronic intercepts, was broadcast on Iranian state television.

The aircraft was shown beneath an Iranian flag, apparently intact after crashing 140 miles inside the country last Sunday. State television claimed that an electronic attack had forced it down by overriding flight systems. American officials acknowledged the loss of the aircraft, but said it was more likely the drone had simply crashed.

Russia and China immediately seized their chance to gain a unique insight into one of the world's most powerful intelligence assets, asking Iran for permission to inspect the drone. Surveillance flights over Iran by CIA-operated aircraft of this kind have been taking place for several years, using bases in neighbouring Afghanistan. Iran's nuclear installations are believed to be the prime target.

The Sentinel, which entered service in 2009, is capable of both intercepting communications and gathering images. No Sentinels are known to have been lost before. Unlike other drones used by the CIA, notably the Predator and the Reaper, the Sentinel carries no weapons and its sole purpose is reconnaissance.

If Russia and China are allowed to inspect the Sentinel, their goal would be to identify — and then replicate — its advanced technology. Elizabeth Quintana, an expert on unmanned air vehicles at the Royal United Services Institute, said they would be particularly interested in the drone's ability to evade detection by radar.

This "stealth technology" is also used by the most advanced "fifth generation" fighters in service with the US Air Force, notably the F-22 Raptor. "The Chinese and the Russians are looking to emulate the US fifth generation fighters," said Ms Quintana. "The Chinese are very keen to replicate any American technologies and they are coming along in leaps and bounds."

They would also want to study the RQ-170's sensors, which allow it to keep large areas under surveillance while the drone loiters, undetected, at an altitude of 50,000 feet. The Russians and Chinese would also focus on its electronic eavesdropping capabilities and the mission systems that allow the aircraft to be guided and controlled.

In 2001, China managed to get hold of a P-3 Orion reconnaissance aircraft, operated by the US Navy, when it was forced down after a mid-air collision. This allowed China to develop counter-measures to the surveillance systems carried by the Orion, forcing the US to upgrade its entire fleet.

Ms Quintana said the American would be worried about the possibility of the "same thing happening" in respect of the Sentinel. It was "standard procedure" for US forces to recover or destroy any drones that are lost, she added. This aircraft appears to have gone down too far inside Iran for any such operation to be feasible.

Russia and China have relatively cordial relations with Tehran, although they allowed the United Nations to pass four resolutions imposing sanctions on Iran.

The Iranian armed forces are likely to use their possession of the drone as a bargaining counter and to seek concessions from Russia and China in return for allowing inspections. But Iran could lack the expertise needed to examine the Sentinel and reverse engineer its technology. Peter Singer, director of the 21st Century Defence Initiative at the

Brookings Institution, said: "What the Iranians are getting from this is symbolic and a shift in the discussion from their nuclear programme and taking over foreign embassies. They are not able to do a lot of reverse engineering on their own, but they have certain nations they have cooperated with in the past. Flights from Moscow and Beijing to Tehran have I'm sure been full this week."

China will also be interested in the drone's engine, added Mr Singer. "The ability to reverse engineer and understand how our aircraft work means the Chinese could build their own derivatives which would be an improvement on what they have now," he said. The Sentinel is believed to fly at 500-600 mph, compared with the 100mph maximum of the better known Predator.

Some American experts suggested that Iran's display of an apparently undamaged drone could have been faked, pointing to the four days that elapsed after its loss. "What I'm looking at is a parade float," said John Pike, director of Global Security.org. "Why did it take so long to show it?"

Drones carry crucial advantages that justify the risk of using them over hostile countries. Satellites orbiting the earth can only cover their targets for relatively short — and predictable — time windows. Drones can loiter for extended periods, beaming back live pictures.

Source: https://www.telegraph.co.uk/news/worldnews/middleeast/iran/8944248/Iran-shows-off-captured-US-drone.html

Saving whales with drones

By KASHMIR HILL
Forbes, 4 January 2012

Melbourne, Australia – When we talk about drones these days, we tend to focus on how law enforcement uses them. The ACLU recently issued a report warning about the possibility of "routine air surveillance of Americans," calling for restrictions on when police can deploy drones. Meanwhile, a robotics philosopher revealed that he was recently called in to brief the CIA's venture capital arm, In-Q-Tel, about the ethics of drones. In an Atlantic piece about the briefing, Patrick Lin mentions that South Korea has drones "that patrol playgrounds for pedophiles." But government agencies are not the only ones finding innovative uses for drones. Activists are starting to get into the drone game. Tim Pool of Occupy Wall Street purchased a camera-equipped Parrot AR drone — that he's named the Occucopter — to make filming confrontations with police easier. On more distant shores, or seas rather, an environmental activist group is using drones to hunt the Japanese whalers hunting whales.

Sea Shepherd is a vigilante environmental organization founded by Paul

Watson (the subject of an excellent *New Yorker* profile a few years back), that tries to prevent Japanese fleets from killing whales in Antarctic waters, by monitoring and ramming them. In order to avoid the Sea Shepherd's boats, Watson says the fleets have started sending harpoon boats to follow his ships when they get close, to relay their coordinates and allow the whalers to escape. But Sea Shepherd gained an advantage in the catfish and mouse game this year by acquiring drones for two of its ships, the *Steve Irwin* and the *Bob Barker*. Osprey drones from Hangar 18 can fly approximately 200 miles from the ships, making it easier to find and keep on the tail of the whalers. The organization bragged that it intercepted the Japanese fleet on Christmas day and again on Wednesday.

"The drones are our eyes in the sky," said Watson, by phone from the *Steve Irwin*. "We absolutely need that long range capability. We also have a helicopter but it's far more expensive to fly, and can't fly in the same wind conditions that the drones can."

A basic Osprey drone costs $8,000, says Jimmy Prouty of Wichita, Kansas-based Hangar 18, which makes drones for hobbyists and for photographers. ("We have a client using a drone to shoot a music video," says Prouty excitedly.) The Sea Shepherd's drones, paid for by a New Jersey recycling company and an anonymous donor, are more expensive deluxe models, outfitted with cameras, GPS, and more robust motors to allow them to fly longer and farther than a basic version.

Watson says his group has had the drones for a year. He says they previously used them in Libya to monitor blue fin tuna poachers.

"We were flying them among the NATO fleet and they didn't even notice them," said Watson. He says the group plans to acquire more drones for its O.R.C.A. Force, and to use them in countries where filming is prohibited, as in Namibia which has banned photography of its annual seal hunt, according to the Examiner.

While the drones seem to have given the Sea Shepherd the advantage for now, it stirred up a bit of trouble for them back home in Australia. When the use of the aerial drones became public in December, the Australian government objected, saying they must undergo an environmental assessment to be used in waters covered by the Antarctic Treaty. The group claims it has since gotten that approval for its Moby Drones.

Source: https://www.forbes.com/sites/kashmirhill/2012/01/04/saving-whales-with-drones/

First man arrested with drone evidence vows to fight case

By JASON KOEBLER
US News, 9 April 2012

LAKOTA, NORTH DAKOTA – The tiny town of Lakota, N.D., is quickly becoming a key testing ground for the legality of the use of unmanned drones by law enforcement after one of its residents became the first American citizen to be arrested with the help of a Predator surveillance drone.

The bizarre case started when six cows wandered onto Rodney Brossart's 3,000 acre farm. Brossart, an alleged anti-government "sovereignist," believed he should have been able to keep the cows, so he and two family members chased police off his land with high powered rifles.

After a 16-hour standoff, the Grand Forks police department SWAT team, armed with a search warrant, used an agreement they've had with Homeland Security for about three years, and called in an unmanned aerial vehicle to pinpoint Brossart's location on the ranch. The SWAT team stormed in and arrested Brossart on charges of terrorizing a sheriff, theft, criminal mischief, and other charges, according to documents.

Brossart says he "had no clue" they used a drone during the standoff until months after his arrest.

"We're not laying over here playing dead on it," says Brossart, who is scheduled to appear in court on April 30. He believes what the SWAT team did was "definitely" illegal. "We're dealing with it, we've got a couple different motions happening in court fighting [the drone use]."

Repeated calls to Brossart's attorney were not returned. Douglas Manbeck, who is representing the state of North Dakota in the case, says the drone was used after warrants were already issued.

"The alleged crimes were already committed long before a drone was even thought of being used," he says. "It was only used to help assure there weren't weapons and to make [the arrest] safer for both the Brossarts and law enforcement."

"I know it's a touchy subject for anyone to feel that drones are in the air watching them, but I don't think there was any misuse in this case," he added.

While there's no precedent for the use of unmanned drones by law enforcement, John Villasenor, an expert on information gathering and drone use with the Washington, D.C.-based Brookings Institution, says he'd be "floored" if the court throws the case out. Using a drone is no different than using a helicopter, he says.

"It may have been the first time a drone was used to make an arrest, but it's certainly not going to be the last," Villasenor says. "I would be very surprised if someone were able to successfully launch a legal challenge [in Brossart's case]."

Villasenor points to two Supreme Court cases — California v. Ciraolo in 1986 and Florida v. Riley in 1989 — that allow law enforcement to use "public navigable airspace, in a physically nonintrusive manner" to gather evidence to make an arrest.

By summertime, there may be many more cases like Brossart's — on May 14, the government must begin issuing permits for drone use by law enforcement.

Currently, about 300 law enforcement agencies and research institutions — including the Grand Forks SWAT team — have "temporary licenses" from the FAA to use drones. Currently, drones are most commonly used by Homeland Security along America's borders.

Bill Macki, head of the Grand Forks SWAT team, says Brossart's case was the first and only time they've used a drone to help make an arrest — they tried one other time (to search for an armed, suicidal individual), but gusty weather conditions made navigation impossible.

With a population of less than 70,000, it doesn't make sense for the Grand Forks police department to own a helicopter, but the ability to call in a drone when necessary can provide a similar purpose.

"The terrain we were working with was very large and agricultural — several hundred acres of very flat farmland made it difficult to set up a perimeter to ensure people didn't make it off the property," he says. "I think drones are definitely a useful tool, their effectiveness in rural operations is exceptional, they keep tactical operations as safe as possible."

Macki is confident his team is trained to legally use drones.

"We've had a relationship with Predator operations for three years, we've provided training for them and received training on the basic capabilities of the Predator," he says. "We've established a relationship with [Homeland Security]. Through that relationship, we've learned drones' capabilities and when we can or cannot use a drone."

Source: https://www.usnews.com/news/articles/2012/04/09/first-man-arrested-with-drone-evidence-vows-to-fight-case

UAV crash in Korea linked to GPS jamming

By CHRIS POCOCK
ainonline.com, 1 June 2012

Incheon, South Korea – Jamming of GPS signals by North Korea may have contributed to the fatal crash of a Schiebel S-100 Camcopter UAV near Incheon, South Korea, on May 10. The small helicopter crashed into its ground control van, killing a Schiebel engineer and injuring the two remote pilots, both Koreans. The jamming started on April 28 and disrupted passenger flights into Seoul's two airports, Kimpo and Incheon. South Korean government officials told local media that the jamming originated from the border town of Kaesong.

Schiebel said that an incorrect response by the operators after the Camcopter lost its GPS signal led to the crash, some minutes later. The UAV is equipped with multiple inertial measurement units (IMUs) for backup, the company noted. The recorders on board the UAV and in the ground station were burned during the crash, and could not provide any explanatory data.

The crashed UAV had been operating in South Korea since October 2008, apparently on demonstrations for government agencies. Schiebel has not identified most of its customers for the S-100. Schiebel lifted a five-day precautionary grounding of the S-100 fleet on May 15.

In an early April briefing attended by AIN, Raytheon UK claimed to be the world leader in GPS anti-jam technology. The company noted that North Korea has been jamming from land vehicles, and that unprotected platforms can be affected at ranges up to 10 miles.

Raytheon UK said it has supplied more than 7,000 jam-resistant antennas to the US and some 20 other countries for aircraft, helicopters, weapons and land vehicles. These include digital versions of the technology that protects platforms over a larger area than the company's earlier, analog system.

Under a Pentagon contract, Raytheon UK has also developed a smaller version suitable for UAVs and light helicopters.

Source: https://www.ainonline.com/aviation-news/defense/2012-06-01/uav-crash-korea-linked-gps-jamming

Pakistan drone attack love song racks up YouTube hits

By JON BOONE
The Guardian, 19 September 2012

ISLAMABAD – In the long history of love songs the attention of a beautiful woman has been compared to many things — but perhaps only in Pakistan's tribal belt would it be likened to the deadly missile strike of a remotely controlled US drone.

In a sign of how the routine hunting down and killing of militants by unmanned CIA planes has leached into the popular imagination, drones have been given a starring role in a new romantic song.

In most respects the track, which is proving popular in the largely Pashtun city of Peshawar, is faithful to standard themes of the genre. The lyrics mention rosebuds and wine. o blaring music it celebrates the allures of a temptress with "sweet lips" and a "smile fresh as early dew" which "ensnares lovers with amorous pangs".

Then the repeated chorus: "My gaze is as fatal as a drone attack".

The hit for singer Sitara Younis follows her success last year with another love ballad, which warns a besotted man to keep his distance: "Don't chase me, I'm an illusion, a suicide bomb."

Khalid Shah Jilani, the professional driver and part-time lyricist who penned the song for a fee of £60, said singers and poets are increasingly taking inspiration from the situation in their provinces.

"It's been a hit because people like the music and the movie that it was written for," he said. "Now you hear it all the time being played at wedding halls and in cars."

Younis's energetic performance of the song had been racking up a healthy number of hits on YouTube before the video sharing site was shut down by the government on Monday, amid rising public anger over the blasphemous film by anti-Islam activists in the US.

Maas Khan Wesal, a Pashtu music veteran who wrote the accompanying music, said the drone reference had nothing to do with politics, but simply the fact that the "eyes of a beautiful dancing girl are so powerful they are like a drone, they can destroy men".

He said the song had proved popular because it reflected the lives of Pashtu speakers on both sides of the Pakistan-Afghanistan border.

"Other people are more political," he said, citing the example of singer Hashmat Sahir. "He sings to tell people that people should not carry out terrorist activities in Peshawar whilst the war in Afghanistan is still going on."

Although music, particularly at once-lively wedding ceremonies, is actively suppressed in areas of the semi-autonomous tribal areas where the Taliban (and drones) are most active, it flourishes in the frontier city of Peshawar.

"In the last five years we have suffered a lot of losses because of this war," said Bakhtiar Khattack, a composer who owns his own recording studio in the city. "People hear about so many different incidents that it becomes part of their psyche."

"It is actually very destructive to our society that these things are being taken lightly and people are even dancing to these sorts of tunes."

The Express Tribune, which reported on the drone love song on Tuesday, said some critics believe such songs are so harmful to Pashtun culture that a board of censors should be set up to stop the material creating "a sense off disunity amongst the people".

The paper said: "All in all, the war on terror in these tribal areas has not only become a part of their art, literature and cinema but even the Pashto tele-films, which are famous for vulgarity and Kalashnikov culture."

Source: https://www.theguardian.com/world/2012/sep/19/pakistan-drone-love-song

Iran claims capture of US drone

By SHIV MALIK
The Guardian, 4 December 2012

TEHRAN – Iran's Revolutionary Guards have captured a US drone flying over the country's airspace, state media reported. The Fars news agency said on Tuesday the Boeing-made ScanEagle drone was gathering information over Gulf waters when it entered Iranian airspace and was subsequently captured by a naval unit of the Revolutionary Guards.

General Ali Fadavi, the Guards' navy chief, was quoted as saying the "intruding" drone was in Iran's possession.

"The US drone, which was conducting a reconnaissance flight and gathering data over the Persian Gulf in the past few days, was captured by the Guards' navy air defence unit as soon as it entered Iranian airspace," Fadavi said. "Such drones usually take off from large warships."

However, the US navy said that none of its drones was missing. "The US navy has fully accounted for all unmanned air vehicles (UAV) operating in the Middle East region. Our operations in the Gulf are confined to internationally recognised water and air space," a US navy spokesman in Bahrain said.

"We have no record that we have lost any ScanEagles recently."

If true, the seizure of the drone would be the third reported incident involving Iran and US drones in the past year.

Last month, Iran claimed a US drone had violated its airspace. The Pentagon said the unmanned Predator aircraft came under fire at least twice but was not hit and that it was over international waters. In April this year, Iran claimed it had copied technology from a US drone brought down in December 2011 on its eastern borders with Afghanistan and Pakistan.

Tehran said it recovered data from the RQ-170 Sentinel, a top-secret drone equipped with stealth technology. As proof, Iranian military cited the drone's flight log, saying it had flown over Osama bin Laden's Pakistani hideout two weeks before he was killed by US special forces.

After acknowledging the loss of the drone, US sources said their software was encrypted and of little intelligence value.

The US blamed the loss of the RQ-170 Sentinel drone on a technical problem, while Iran claimed it brought it down electronically by disrupting its GPS system.

Source: https://www.theguardian.com/world/2012/dec/04/iran-claims-capture-us-drone

WWF plans to use drones to protect wildlife

By ADAM VAUGHAN
The Guardian, 7 February 2013

AFRICA/ASIA – Conservation group WWF has announced plans to deploy surveillance drones to aid its efforts to protect species in the wild, as the South African government revealed that 82 rhinos had been poached there since the new year.

The green group says that by the end of the year, it will have deployed "eyes in the sky" in one country in Africa or Asia, with a second country following in 2014 as part of a $5m hi-tech push to combat the illegal wildlife trade.

A record 668 rhinos were killed by poachers in South Africa alone last year, and a single shipment of ivory seized in Malaysia in December weighed almost as much as all the illegally traded ivory since in 2011, which was itself a record year for seizures. And poachers have kept up their hit rate since the beginning of 2013, according to figures from the South African government. "The Kruger National Park remains the hardest hit by rhino poachers this year, having lost 61 rhino to mostly foreign poachers," a government spokesperson said. "Twenty-one poachers have been arrested, 14 of them in the Kruger National Park."

The criminal trade has become so serious that last year the US intelligence community were ordered to track poachers by then secretary of state Hillary Clinton, with a WWF report in December warning the multibillion dollar trade was now threatening national security in some countries.

WWF's three-year project also involves combining data from unmanned aerial vehicles, cheap mobile phone technology tracking animal movements, and handheld devices carried by rangers, in a bid to outsmart often heavily armed poachers who bribe corrupt officials to avoid patrols and find wildlife.

Allan Crawford, project leader for the WWF Google technology project, who had just returned from the Kruger national park where many of South Africa's rhinos are being killed, told the Guardian: "It's a very scary prospect for rangers … they could run into very heavily armed gangs of poachers, there's usually four or five of them, sometimes with dogs. They've also got wild animals to contend with — one ranger was recently attacked by a lion. They're outnumbered, and sometimes poachers have night-vision equipment. There aren't enough resources to tackle this in South Africa at the moment. This is where the new technologies come in, to help them."

Drones are already being used by conservationists to monitor wildlife, such as orangutan populations in Sumatra, anti-whaling activists are using them against the Japanese whaling fleet, and a charity in Kenya recently beat its target of raising $35,000 in crowdfunding for a drone to protect rhinos and other wildlife in the country's Laikipia district. One South African rhino farmer is even planning to put 30 drones in the sky himself. But the way the three key technologies are being used by WWF is "unprecedented", Crawford said.

A pair of drones will be used in each of the two countries selected, which the group hopes to name within weeks, with plans to ultimately be operational in four sites by 2015, with different terrains. Crawford said the software and drones, which would be operated by rangers or local law enforcement, would "generate a strategic deployment of rangers in the most cost effective way, so they can form a shield between animals and poachers."

The drones would likely cost in the tens of thousands of dollars rather than hundreds of thousands, he said, adding that "if governments wanted to, they could deploy [more expensive] high altitude drones that can stay up high in the sky, and track poachers to get the middle men and whole trade line." The funding for WWF's project comes from a Google grant awarded last December.

"We've got to crack this problem because it's getting out of control," Crawford said. "The poachers seem to have figured out how to get round existing anti-poaching methods." But he admitted anti-poaching efforts could get caught up in an arms race, and that it was a "risk" that poachers may get their own drones. "It will be an escalation like that until we change the cause, which is demand in Asia."

Source: https://www.theguardian.com/environment/2013/feb/07/wwf-wildlife-drones-illegal-trade

Ethiopia produces first military drone aircraft

By TESFA-ALEM TEKLE
Sudan Tribune, 14 February 2013

Addis Ababa – An Ethiopian military source has told *Sudan Tribune* that the country has built the first unmanned aerial vehicle (UAV) or drone which could be used for multiple purposes.

After undergoing testing, the locally made drones, have demonstrated their capability of performing a number of militarily and civilian applications, according to the source.

Speaking on condition of anonymity from the country's air force base in Debrezeit town, a military official told *Sudan Tribune* that the drones are equipped with onboard sensors, cameras and GPS to carry out cost-effective monitoring activities even across difficult landscapes like the highlands of Ethiopia.

Besides serving in a number of military missions — such as in monitoring border security — the UAVs will also be deployed to perform geophysical surveys, assist forest protection and monitor forest fires or other natural disasters.

The drones have already made test flights performing a geophysical survey of Ethiopia's controversial grand renaissance dam, a massive hydro-power plant project the country is constructing on the Blue Nile River near to the Sudanese border.

In recent years, many African countries have shown growing interest in using drones as a cost-effective way to control huge infrastructure facilities, as well as areas rich in natural resources such as oil, mine and gas sites.

In 2011 Ethiopia signed an agreement with Israeli manufacturer BlueBird Aero Systems to purchase drones.

Binyam Tekle, a lecturer and researcher at a government university, says the development of indigenous drones is a great achievement for Ethiopia and will help strengthen the national army.

Due to Ethiopia's long and fragile borders with Eritrea, Somalia, Kenya, Sudan and more recently South Sudan, he said it is timely for the country to use UAVs to monitor these shared and often tense and porous zones.

"With Eritrea-backed rebels and Somalia's *al-Qaeda* linked *al-Shabaab* terrorists repeatedly posing threats to national security, using UAVs will be crucial for Ethiopia to avert planned attacks," he told *Sudan Tribune*.

Ethiopia is a key regional security partner to the United States particularly in the war on terror due to its proximity to Yemen and Somalia.

In 2011, the Obama administration launched a drone base in Ethiopia for counter-terrorism operations in the Horn of Africa, particularly to attack *al-Qaeda* affiliates in Somalia. Earlier this month, it was revealed that the US has had a drone base in Saudi Arabia, with its existence kept secret by the US media in collusion with the Obama administration.

In recent years, Ethiopia has made tremendous achievements in the defence sector by managing to manufacture its own military equipment and defence systems.

On Thursday, Ethiopia marked its first ever Defence Force Day under the theme "Our constitutional loyalty and public nature would be preserved".

A defence exhibition was staged in the heart of the capital, Addis Ababa, demonstrating the level of progress the nation has made.

Light and heavy modern weapons, as well as different vehicles manufactured by the army-run automotive industry were also displayed at the exhibition.

Government officials said that Ethiopia has built a defence force capable of breaking any internal or external enemy.

The Horn of Africa nation has one of the strongest army and air forces on the continent and often contributes troops to United Nations peace keeping missions.

Ethiopia spends around 2.4% of its GDP on the military.

> *Source: http://www.sudantribune.com/spip.php?article45518*

Drones should be banned from private use, says Google's Eric Schmidt

By JAMES BALL
The Guardian, 21 April 2013

MOUNTAIN VIEW, CALIFORNIA – The use of cheap, miniature "everyman" drones needs to be banned by international treaties before such devices fall into the hands of private users including terrorists, the head of Google has said.

In an extended interview with the Guardian, Eric Schmidt, executive chairman of Google and an adviser to Barack Obama's presidential campaign, warned of the potential of new technology to "democratise the ability to fight war", and said drones could soon be used to harass and spy on neighbours.

"You're having a dispute with your neighbour," he hypothesised. "How would you feel if your neighbour went over and bought a commercial observation drone that they can launch from their back yard. It just flies over your house all day. How would you feel about it?"

Schmidt set out the trajectory of robotic warfare and considered whether it would be confined solely to national governments. "It's probable that robotics becomes a significant component of nation state warfare," he said.

"I'm not going to pass judgment on whether armies should exist, but I would prefer to not spread and democratise the ability to fight war to every single human being."

"It's got to be regulated. You just can't imagine that British people would allow this sort of thing, and I can't imagine

American people would allow this sort of thing. It's one thing for governments, who have some legitimacy in what they're doing, but have other people doing it … It's not going to happen."

The US government's use of military drones has proven increasingly controversial, with drone strikes on American citizens the subject of a recent 12-hour Senate filibuster by the Republican senator Rand Paul. The Bureau of Investigative Journalism estimates that US drone strikes in Pakistan, Yemen and Somalia have been responsible for at least 2,772 deaths.

Schmidt's warnings on privacy in the robotic era notwithstanding, Google itself has been frequently criticised by privacy campaigners concerned about the company's huge reach and the extensive data collection used to power its multi-billion-dollar advertising sales.

Challenged on these issues, Schmidt said Google was "super-sensitive" on privacy and had voluntarily kiboshed projects it thought could lead to privacy breaches. "Google is not a bunch of engineers who throw stuff over the wall," he said. "A classic example is that a team built a facial-recognition tool. It was just really good — state of the art at the time. We stopped that product for two reasons. One is that it turned out to be illegal in Europe and the second was that it was not a good product to offer in the US for the same reasons."

Schmidt, who said he was "literally in the room" when the decision was made to kill the product, said it had been a judgment call taken on Google's own initiative.

"Facial recognition, completely unmonitored, can be used for very bad things," he said. "It can be used for stalking, for example. You know, it's just we don't want to be part of that as a company. There are cases where facial recognition can be used, but they need to be fairly carefully boxed."

Schmidt also addressed the "transition fund", valued at between €50m and €60m (£33m-£39m), set up by Google after negotiations with the French government. The fund will support technologies to help French publishing companies that are suffering during the transition to digital to monetise their content.

Schmidt avoided the question of whether a similar fund could be established in the UK. "I'm sure we can talk about it," he said. "The reason I like this model is it's … I don't like the idea of randomly writing cheques to publishers in the old model, and I think it's a very good idea for Google to assist in the transformation of their business model from old to new."

Source: https://www.theguardian.com/technology/2013/apr/21/drones-google-eric-schmidt

Canadian police save a man's life, using a drone

By Ben Coxworth
New Atlas, 13 May 2013

SASKATOON – While we hear a lot about the ways in which hovering aerial drones can potentially be used to violate peoples' privacy, it's always nice to know that they can help us, too. That was the case last Thursday (May 9th), when RCMP from the Canadian city of Saskatoon, Saskatchewan used a remotely-operated quadcopter to locate the victim of a single-vehicle rollover, which occurred in the countryside at near-freezing temperatures.

At 12:20 am, the Saskatoon RCMP first received a report of the rollover, and went out to the accident site to investigate — they were joined at the scene by fire rescue and emergency medical response teams. The car was located, but even after conducting a 200-meter (656-foot) ground search of the area, no occupants could be found.

An air ambulance helicopter was then called in, the crew of which used a night vision system to conduct a larger, aerial search. When that search also proved fruitless, the RCMP brought in Corporal Doug Green, a Forensic Collision Reconstructionist, along with the detachment's forward-looking infrared (FLIR) camera-equipped Draganflyer X4-ES quadcopter — the drone is made by Draganfly Innovations, which is itself based in Saskatoon.

Source: https://newatlas.com/rcmp-quadcopter-locate-victim/27488/

Delivery drones are coming: Jeff Bezos promises half-hour shipping with Amazon Prime Air

By DAVID PIERCE
The Verge, 1 December 2013

SEATTLE – Jeff Bezos is nothing if not a showman. Amazon's CEO loves a good reveal, and took the opportunity afforded by a *60 Minutes* segment to show off his company's latest creation: drones that can deliver packages up to 5 pounds to your house in less than half an hour. They're technically octocopters, as part of a program called "Amazon Prime Air."

A drone sits at the end of a conveyer belt, waiting to pick up a package — Bezos says 86 percent of Amazon's packages are under 5 pounds — and can carry them up to 10 miles from the fulfillment center. As soon as Amazon can work out the regulations and figure out how to prevent your packages from being dropped on your head from above, Bezos promised, there will be a fleet of shipping drones taking the sky.

The segment focused primarily on holiday shopping, particularly the annual shopping extravaganza that is Cyber Monday.

It's a huge day for Amazon — more than 300 items will be ordered each second — and does much to reveal the company's true ambitions. Amazon doesn't just want to reinvent the way we shop for and buy things, it wants to upend every step of the process — including how our purchases come to us.

Charlie Rose spoke to a number of Amazon executives and employees, and toured one of the company's 96 massive warehouses, known as "fulfillment centers." Its burgeoning same-day delivery infrastructure was on full display and Rose explored the company's sprawling moves into fashion, groceries, web hosting, tablets, and strange political shows starring John Goodman.

But Jeff Bezos was predictably the star of the segment. He talked about how, 18 years ago, he'd drive packages to the post office himself and dreamed of one day owning a forklift. (Things have changed a bit since then.) He says we're four or five years from drones being able to deliver small packages right to your house, largely because the company has to work with the FAA to make sure it's legally allowed to run the Prime Air program — Amazon doesn't have Zookal's luxury of operating in Australia without the FAA's regulatory oversight.

Bezos knows there's work to do, but his overwhelming message was simple: get ready. "It will work, and it will happen, and it's going to be a lot of fun."

Source: https://www.theverge.com/2013/12/1/5164340/delivery-drones-are-coming-jeff-bezos-previews-half-hour-shipping

UN launches spy drones to pressure rebels in Democratic Republic of Congo

By DAVID SMITH
The Guardian, 4 December 2013

KINSHASA – The UN is using surveillance drones for the first time in its history in an attempt to step up the pressure on militias in the Democratic Republic of the Congo.

The first two spies in the sky were launched on Tuesday from the eastern city of Goma, which last year was briefly occupied by the M23 rebel group. UN officials expect another drone to be deployed by the end of the month and two more by the end of March. They have a range of 155 miles from their base in Goma.

Presiding over an unveiling ceremony at a rickety airport, Hervé Ladsous, the UN peacekeeping chief, described the unmanned aerial vehicles as "an essential instrument ... to observe armed groups and also to keep watch over civilians who are displaced or on the run due to the conflict".

The Congolese defence minister, Luba Tambo, said the drones would play a critical role in helping patrol the nation's porous border with Rwanda, which was accused of sending fighters and weapons to M23 rebels.

The M23 was defeated by the Congolese army backed by a UN intervention brigade last month. But the UN has said it will turn its attention to other armed groups in the region. Among them are the Democratic Forces for the Liberation of Rwanda (FDLR), led by extremist Hutus from Rwanda who took part in that country's 1994 genocide, and the Allied Democratic Forces (ADF), an Islamist militant group.

"We need to take care of the FDLR, the ADF, the Mai Mai," Ladsous told the Associated Press. "We are going to use these machines and they will have an important deterring effect."

If they prove successful in Congo, he added, they could also be used in other UN peacekeeping missions. Commanders in South Sudan have reportedly expressed an interest, while Mali could be another candidate.

In January, the UN security council approved the trial use of unarmed drones for intelligence gathering in eastern Congo. It has chosen a model known as the Falco produced by Selex ES, a subsidiary of the Italian giant Finmeccanica. The Falco is capable of carrying several types of high-resolution sensors.

Alberto Dos Santos Cruz, the UN force commander in Congo, told the BBC that the drones would only fly over Congolese territory, as they have no mandate to operate in neighbouring countries.

Congo has been engulfed in conflict for nearly two decades and armed groups have thrived in its mineral-rich east, despite the presence of nearly 20,000 blue-helmeted UN peacekeepers.

Source: https://www.theguardian.com/world/2013/dec/04/un-spy-drones-in-congo

Jonathan unveils Nigerian-made drone

By KINGSLEY OMONOBI
Vanguard Nigeria, 18 December 2013

ABUJA, NIGERIA – President Goodluck Jonathan, yesterday, unveiled the first indigenous unmanned aerial vehicle, UAV, named GULMA, built by the Nigerian Air Force in Kaduna.

He declared that the giant technological breakthrough of various sectors, particularly the Armed Forces, are pointers to the country's potential for greater deeds.

In his speech at the unveiling ceremony, President Jonathan said: "We must not forget the lesson of this encouraging moment in our nation's history; that as Nigerians we are capable of great deeds."

"In the same manner, we must not forget that ahead of us still lie challenges that we must surmount. We must, therefore, stay focused on our common national purpose."

"Today marks another landmark moment in our nation's history, as we gather here to witness the unveiling of our country's first indigenous unmanned aerial vehicle, which has been designed and constructed by the Nigeria Air Force."

"Besides its diverse military application, unmanned aerial vehicle provides mankind with a range of benefits in disaster management, power line surveys, law enforcement, telecommunications, weather monitoring, and aerial imaging/mapping. It is also rapidly becoming an important tool in news coverage, environmental monitoring, and oil and gas exploration."

"Considering the potential impact of its benefit and versatility, I cannot but say how proud I am of the men and women of our Armed Forces."

"Apart from their commitment to the protection of our sovereignty, they are helping to keep our nation ahead in military science and technology and to keep their civilian counterparts on their toes."

Minister's comments

The Minister of Information, who also supervises the Defence Ministry, Labaran Maku, noted that the true professionalism of the military since 1999 had given rise to the indigenous military-led technological breakthroughs.

Maku said: "The development of surveillance aircraft by the Nigerian Air Force at this crucial moment in our history is a bold statement of our resolve to provide security for the Nigerian populace."

"The GULMA UAV potentially affords the Armed Forces a window of immense opportunities in its task of acquiring timely intelligence and combating criminality, especially in light of the nation's prevailing security challenges."

Gulma's history

In his speech, Air Marshal Alex Badeh explained that the surveillance aircraft was named GULMA, a word in Hausa that means gossip, because of its function as an intelligence, surveillance and reconnaissance platform.

According to him, the idea began in 2005 with the acquisition of three Aerostar UAVs for the defunct Presidential Committee on Maritime Safety and Security, PICOMSS, which was frustrated "by numerous vendor-related problems that eventually led to the grounding of all the three systems in 2009."

"Following the dissolution of PICOMSS in 2012, the Federal Government handed over the UAVs to the NAF." He said NAF Optimising Local Engineering, NAF OLE, which handled the project, did it at the cost of $3 million, half of the $6 million quoted by foreign vendors and also trained 15 UAV pilots at less than half a million dollars compared to the $2 million quoted by foreign vendors to train only three pilots.

Source: https://www.vanguardngr.com/2013/12/jonathan-unveils-nigerian-made-drone/

0016
RONALD
REAGAN

People burn a mock US flag during a protest against drone attacks in Pakistan. ©STR/EPA

The Japanese whaling ship Nisshin Maru hunting in Antarctic waters as seen from the Sea Shepherd drone. ©Sea Shepherd

6002

Amazon Prime Air will be able to deliver packages up to five pounds in 30 minutes or less. © Amazon

Woman athlete suffers head injuries after 'hackers took control of drone and made it crash'

By LILLIAN RADULOVA
Daily Mail, 8 April 2014

GERALDTON, AUSTRALIA – An Australian athlete suffered head injuries after a a drone which had been filming the race plummeted to the ground.

Raija Ogden was only meters from the finish line in the Endure Batavia Triathlon in Geraldton, Western Australia, on Sunday, when the incident happened. She claims the drone hit her on the head and that an ambulance crew later found a piece of propeller embedded in her skull. She was taken to hospital where she needed three stitches.

The drone owner and operator Warren Abrams said Mrs Ogden was hurt when she fell over after being startled by the piece of equipment as it fell. The Civil Aviation Safety Authority (CASA) is investigating.

"My hair was completely red with blood." Mrs Ogden told The Western Australian. 'I didn't hit the ground. I sat down because I just thought I was going to pass out.' She vehemently denied that the injuries were caused when she fell over. Warren Abrams, the pilot and drone owner from New Era Photography and Film, admitted that he had lost control of the drone before it fell 10 meters to the ground.

However, he claimed that this was because it had been hacked and that he had also lost control in an earlier test flight when someone 'channel hopped' and took it over. Mr Abrams believes any of the spectators or competitors could have easily used a smartphone to hack into the drones operating system. He also claimed that footage from the flying camera would show that the device missed Ms Ogden and fell just behind her.

A number of witnesses have expressed varying views, supporting both versions of the event.

CASA is looking into the incident concerning the regulations for operating the drones, which dictate the minimum distance that unmanned aircraft must be flown is 30 metres away from people at all times. The regulations relating to commercial unmanned aircraft clearly state that they 'require UAV controller's certification and an operator's certificate to fly.'

Source: http://www.dailymail.co.uk/news/article-2599269/Australian-triathlete-injured-crashing-drone-pilot-loses-control.html

Pepper-spray drone offered to South-African mines for strike control

By DAVID SMITH
The Guardian, 20 June 2014

JOHANNESBURG – A South African company has built a drone designed to shower pepper spray on unruly crowds and says it has begun supplying units to an international mining company.

Desert Wolf claims it wants to help in "preventing another Marikana" — a reference to a protest in August 2012 at which 34 striking mineworkers were shot and killed during clashes with the police.

But the "Skunk Riot Control Copter" was condemned by labour activists as "absolutely outrageous" and compared with deadly US military drones in Pakistan.

Desert Wolf is marketing the 500,000 rand (£27,400) machine, unveiled at a recent trade show, as "designed to control unruly crowds without endangering the lives of the protesters or the security staff".

It says the Skunk boastscan eight electric motors with 16-inch propellors, lifting 45kg and carrying 4,000 pepper-spray paintballs, plastic balls or other "non-lethal" ammunition. The device is equipped with four barrels firing up to 20 balls per second each, which could equate to 80 pepper balls per second "stopping any crowd in its tracks".

The firm's website states: "The operator has full control over each marker. He can select the RED paint marker and mark the protester who carries dangerous weapons, he can select the BLUE marker to mark the vandalising protestors and if needed the pepper balls to stop the advancing crowd before they get into a 'life threatening situation'."

The Skunk is also fitted with bright strobe lights, "blinding lasers" and on-board speakers to enable communication and warnings to the crowd. It has a thermal camera and high-definition video camera with on-board recording.

Desert Wolf insists: "Our aim is to assist in preventing another Marikana; we were there and it should never happen again."

On a single afternoon in Marikana, police gunned down 34 striking mineworkers in the bloodiest act by security forces since the end of racial apartheid in 1994. The killings are the subject of an ongoing commission of inquiry. Ten others, including police and security guards died in clashes that week.

Hennie Kieser, managing director of Desert Wolf, said he was flying drones at Marikana at the time as part of the surveillance operation.

"Anyone who was at Marikana would rather have this technology than live ammunition," he continued. "People who say it's inhumane compared to 9mm bullets are idiotic."

Kieser insisted: "It's the best solution for both parties: the miners on strike and those who are trying to control the strike. They both want to go home to their wives at the end of the day."

He said an international mining house had ordered 25 of the unmanned aerial vehicles but he is not permitted to disclose its identity, or say where it operates. There is also interest from mining and security companies in South Africa and beyond, he said. Current aviation rules in South African prevent the drones being deployed but he hopes the legislation can be changed.

There was an immediate backlash against the technology. James Nichol, a British lawyer representing the families of dead strikers at Marikana, said: "It's absolutely outrageous. Using pepper spray like ammunition to scatter the crowd. People are entitled to be on strike. Who would make the decision? It's absurd."

He added: "What we know about drones is in Pakistan they have killed funeral parties and they have killed wedding parties. Innocent people would be caught up in this. It seems to be the thin end of the wedge."

"One of the lessons of Marikana is that the state should stay out of industrial disputes. If they want people firing back at drones which then crash and hurt innocent people, be it on their heads. It's disgraceful."

Rehad Desai, spokesperson for the Marikana Support Campaign and director of a documentary film, Miners Shot Down, said: "The government are increasingly turning to authoritarian methods instead of dialogue and mediation. It's to be expected that they would adopt such equipment to quell dissent. But the more violent the equipment, the more violent the reaction will be."

"Why would it be more 'humane' to pepper spray people on their way home than to shoot them? I don't see the analogy."

Tim Noonan, a spokesman for the International Trade Union Confederation, told the BBC: "This is a deeply disturbing and repugnant development and we are convinced that any reasonable government will move quickly to stop the deployment of advanced battlefield technology on workers or indeed the public involved in legitimate protests and demonstrations."

South Africa's platinum belt is currently deadlocked in a strike which, after five months, is the longest in the history of the country's mines and has pushed many workers and their families into dire poverty and hunger.

Source: https://www.theguardian.com/world/2014/jun/20/pepper-spray-drone-offered-south-african-mines-strike-control

Dutch artists turn boy's dead pet into rat-copter

By SARAH CASCONE
ArtNet News, 11 September 2014

EDE, NETHERLANDS – Arjen Beltman and Bart Jansen have finally topped *Orvillecopter*, their 2012 cat drone, creepily created from the body of Jansen's cat Orville, who died in a car accident. The latest from the Dutch artist-inventor duo is a rat-copter, which transforms and memorializes a Dutch schoolboy's pet rat, reports the BBC and Daily Mirror.

The rat, named Ratjetoe, the Dutch word for ratatouille, had to be put to sleep after being diagnosed with cancer. His owner, 13-year-old Pepijn Bruins, was heartbroken to lose his pet, saying in an interview that "when I learned he had cancer and the vet had to put him to sleep I was very upset. I had seen Bart and Arjen and their flying cat, and I asked my dad if it would be possible to have the rat fly."

Luckily for the boy, the artists were more than up to the unusual task of turning his deceased pet into a flying machine. "When I heard how upset he was, I just knew I had to help," Beltman said in an interview. "Technically, it was very different to the cat, it has three rotors instead of four, and being a small rodent, it is extremely lightweight, so prone to being blown by the wind." Mounted on a specially rigged flying device, Ratjetoe is now capable of soaring through the air thanks to three noisy, bright red propellers.

While preserving a dead rat may seem creepy to some, it's made the loss of a beloved pet that much easier to take for Bruins. "I loved him very much," the boy said. "He always liked to be cuddled and he would run up my clothes and hide." And, as Jansen points out, "flight is man's greatest achievement, so why not give it to more animals? The world needs more flying animals."

Since making headlines with *Orvillecopter*, Beltman and Jansen have also created *Turbo Shark* — which sounds like a prop Damien Hirst might design for the inevitable third *Sharknado* film — and enabled the famously flightless ostrich to take wing. Their work was recently featured in *All Creatures Great and Stuffed*, a television documentary that debuted on the BBC's Channel 4 yesterday, September 10.

Source: https://news.artnet.com/art-world/dutch-artists-turn-boys-dead-pet-into-rat-copter-100136

Argentina uses drones to root out wealthy tax evaders

By NICK ALLEN
The Telegraph, 25 September 2014

BUENOS AIRES – The Argentine government has used drones to catch out wealthy tax evaders who had not declared mansions and swimming pools.

Unmanned aircraft were dispatched over an upper class area of Buenos Aires and discovered 200 homes and 100 pools that had not been detailed on returns.

Tax officials said the drones took pictures of luxury houses standing on lots registered as empty.

The evasions found by the drones amounted to missing tax payments of more than $2 million and owners of the properties have been warned they now face large fines.

The unregistered mansions and pools were found in an area about 10 miles south of the city and, according to the tax agency, they appeared to be large and constructed with "premium materials".

Use of drones has been expanding in Argentina and the rest of South America with the unmanned vehicles being deployed for purposes as diverse as locating drug smuggling routes, monitoring farm crops, and looking for archaeological sites.

According to the Council on Hemispheric Affairs, most of the drone technology being sold to South American governments and companies — $500 million worth between 2005 and 2012 — comes from Israel, but drones are also being produced cheaply in Mexico.

The proliferation has raised concerns among human rights activists.

Speaking to South America politics magazine Upside Down World last year, Alejandro Sanchez, senior research fellow at the Council on Hemispheric Affairs, said: "Anyone thinking drones are financially unattainable for less developed countries hasn't looked at the latest models."

"There's definitely a need for a technology that's both cheap and can have some really positive results but obviously there's a possibility this technology can be used for all the wrong reasons and, unfortunately, throughout Latin America's history, the abuse of power tends to happen quite often."

> *Source: http://www.telegraph.co.uk/news/worldnews/southamerica/argentina/11121984/Argentina-uses-drones-to-root-out-wealthy-tax-evaders.html*

Serbia vs Albania abandoned after players and fans brawl on pitch

By NICK AMES & SASA IBRULJ
The Guardian, 14 October 2014

BELGRADE – Serbia's Euro 2016 qualifying match against Albania in Belgrade was abandoned with the score goalless after fighting broke out between players, fans and team officials on the pitch of the Partizan Stadium.

The brawl, which involved representatives of both sides and a number of pitch invaders from the stands, arose after a drone had been flown above the pitch, lowering a flag bearing an insignia of 'Greater Albania'. The flag was caught by the Serbia defender Stefan Mitrovic, setting off chain of events that is likely to lead to severe repercussions from UEFA — who will now consider the matter.

The match was Albania's first visit to Belgrade since 1967 and a highly sensitive affair given long-running tensions over Kosovo, which declared independence from Serbia in 2008 and has a majority ethnic Albanian population. There had been little sign of what was to come in the early stages, with away supporters barred from attending the fixture and the home side controlling the early play against a defensively-minded visiting team.

In the 40th minute, the English referee Martin Atkinson halted play — for a second time — after flares had been thrown from the crowd in the direction of Albanian players. At the same time, a drone — seemingly remote-controlled — was seen hovering above the pitch, parallel to the top of the stand, before falling towards the pitch 15 yards into the Serbian half. The 'Greater Albania' insignia referred to the notion of an extended area in which all ethnic Albanians reside — one that would include Kosovo.

Mitrovic caught the flag and was immediately accosted by Albanian players Andi Lila and Taulant Xhaka, the insignia eventually ending up in the hands of the striker Bekim Balaj, who began to carry it towards the touchline. By this time, officials and substitutes from both teams had entered the pitch, the Serbian replacement Nenad Tomovic running from the bench and blocking Balaj's path. At the same time, a supporter attacked Balaj with a plastic chair which prompted a violent melee, the Albania captain Lorik Cana chief among the protagonists as he held the fan to the ground.

The Albanian players ran towards the players' tunnel, ducking for cover amid a shower of missiles and having to take evasive action as at least one supporter attempted to assault a visiting player. They were followed by the Serbs as play was suspended for about 40 minutes, before the game was finally abandoned.

Serbian media have claimed that the drone was controlled by Orfi Rama, brother of the Albanian prime minister Edi Rama, from the VIP section. There are also unconfirmed reports that he was arrested at the stadium. The uefa match delegate, Harry M Been, said: "We regret the situation. You all saw what happened and I cannot comment on who is to blame or what to blame. I will submit a report with my colleagues to UEFA and UEFA will decide what will happen further."

Serbia captain and Chelsea defender Branislav Ivanovic added: "We wanted to continue the match, but Albanian players said they were not psychologically ready."

Source: https://www.theguardian.com/football/2014/oct/14/serbia-albania-euro-2016-flag-halted

Ambulance drone delivers help to heart attack victims

By MICHELLE STARR
CNET, 28 October 2014

Delft, Netherlands – An ambulance drone carrying a defibrillator for rapid response to heart attacks has just been unveiled in the Netherlands.

One of the most important considerations in emergency medical treatment is response time. Now, an engineering graduate at Delft University of Technology has created a rapid response drone to do what a regular ambulance can't. The drone, created by Alec Momont, is able to fly at speeds of up to 100 km/h (60 mp/h), carrying a defibrillator and equipped with features that could reduce the time before a heart attack victim receives first aid, greatly increasing the chances of recovery.

"It is essential that the right medical care is provided within the first few minutes of a cardiac arrest," Momont said. "If we can get to an emergency scene faster, we can save many lives and facilitate the recovery of many patients. This especially applies to emergencies such as heart failure, drownings, traumas and respiratory problems, and it has become possible because life-saving technologies, such as a defibrillator, can now be designed small enough to be transported by a drone."

The prototype drone is designed to be deployed when emergency services receive a cardiac arrest call. Unconstrained by traffic and roads, the drone, in theory, could arrive at the scene faster than an ambulance. Because it cannot, however, carry EMTs, it is equipped with the next best thing: livestream audio and video connection that will allow medical professionals to deliver instructions to people at the site, viewing the situation through the webcam and talking the responder through the treatment — including how to use the defibrillator.

Currently, only 20 percent of untrained people are able to use a defibrillator with any success; with an emergency technician issuing instructions via webcam, this could be increased to 90 percent, Momont said.

"Some 800,000 people suffer a cardiac arrest in the EU every year, and only 8 percent survive," he said. "The main reason for this is the relatively long response time of the emergency services (approx. 10 minutes), while brain death and fatalities occur within four to six minutes. The ambulance drone can get a defibrillator to a patient inside a 12 km^2 zone within one minute. This response speed increases the chance of survival following a cardiac arrest from eight percent to 80 percent."

At this point in time, there are some barriers to a widespread deployment of the ambulance drone. Air traffic laws in the Netherlands forbid the use of autonomous drones; this legislation is expected to be rectified sometime in 2015.

The drone is also yet to be tested on real patients, and the system of object detection and avoidance in the drone needs to be refined. However, Momont — who developed the drone in collaboration with Belgian innovation platform Living Tomorrow — believes that it could be as little as five years before his drone is in use. Several medical sector bodies have also registered interest in the project.

"The costs should not be an issue; I have calculated these at approximately €15,000 per drone, which is clearly a reasonable amount if you consider the number of lives that could be saved," Momont said. "I hope it will save hundreds of lives in the next five years."

Source: http://www.cnet.com/news/ambulance-drone-delivers-help-to-heart-attack-victims/

China unveils laser drone defence system

By AGENCE FRANCE PRESSE
The Guardian, 3 November 2014

BEIJING – China has developed a highly accurate laser weapon system that can shoot down light drones at low altitude, state media reported.

The machine has a 1.2-mile range and can bring down "various small aircraft" within five seconds of locating its target, the official Xinhua news agency said, citing a statement by the China Academy of Engineering Physics (CAEP), one of the developers.

Xinhua showed pictures of large metal boxes in camouflage paint and the wreckage of a small drone, some of it burning.

The laser system is expected to "play a key role in ensuring security during major events in urban areas" and address concerns on unlicensed mapping activities, according to Xinhua. It is effective up to a maximum altitude of 500 metres and against aircraft flying at up to 50 metres per second (112mph), Xinhua said.

It cited Yi Jinsong, a manager with China Jiuyuan Hi-Tech Equipment Corp, as saying that small-scale, unmanned drones were relatively cheap and easy to use, making them a likely choice for terrorists.

"Intercepting such drones is usually the work of snipers and helicopters, but their success rate is not as high and mistakes with accuracy can result in unwanted damage," he said.

The system, which can be installed on vehicles, shot down more than 30 drones in a recent test with a "100% success rate" said the CAEP statement.

The academy is developing similar laser security systems with greater power and range, Xinhua said in its report on Sunday.

China typically deploys tight security for domestic political meetings, international conferences and sports events, looking to prevent public protests over issues such as illegal land seizures and corruption as well as any threats to the participants.

> *Source: https://www.theguardian.com/world/2014/nov/03/china-unveils-laser-drone-defence-system*

French government on alert after unexplained drone flights over nuclear power stations

By JOHN LICHFIELD
The Independent, 9 November 2014

PARIS – An epidemic of mysterious — and potentially disturbing — drone flights over French nuclear power stations remains unexplained despite the recent arrests of three young model aircraft enthusiasts in central France.

The illegal flights by the tiny, pilotless helicopters, mostly at night, were initially dismissed as a nuisance. But a recent spate of five co-ordinated "visits" in one evening to nuclear reactors hundreds of miles apart has now placed the French government on high alert.

A campaign of harassment by anti-nuclear campaigners is considered the most likely explanation. Surveillance flights by a terrorist group testing the security of France's 19 nuclear sites have not been ruled out. Then, last Wednesday, three people, two men aged 24 and 31 and a woman of 21, were arrested close to a power station at Belleville-sur-Loire in Cher. Police said the three were about to launch a relatively simple drone — a type sold on the internet for around €100.

They face possible charges but are not suspected of being responsible for the score of intrusions by much more elaborate drones in the restricted airspace over 13 nuclear power stations since early October. "They appear to be model aircraft enthusiasts," a source in the investigation told the newspaper Le Parisien. "The machine they were using was sophisticated enough but not military or professional. It was a toy. Whether they were doing it for fun or they had some political motive is not yet clear."

Either way, the arrested three are believed to be copycats rather than members of the team which has organised the elaborate drone flights.

An emergency inter-ministerial meeting has considered several possible explanations for the flights, ranging from the irritating to the alarming. This followed five drone incursions in the space of six hours at nuclear sites ranging from the Channel coast in Normandy to the Franco-German border in Alsace. The hundreds of miles between the flights suggested elaborate advance planning by a group of several people.

After the meeting last week, some government sources said the special gendarmerie teams which guard nuclear sites had been given authority to shoot down future intruders. Three more flights followed. No drones have been shot down so far.

Many of the flights have taken place at night but the authorities have filmed and photographed some of the intrusions. Officials say they have been surprised, and alarmed, by what the images have revealed. They are not the "toy" drones that can be easily acquired and operated from a smartphone. Up to 400,000 such drones are believed to exist in France. The intruders are something much more complex and expensive — helicopter-like drones with powerful engines capable of flying for dozens of miles. They have a triangle of three white lights, a larger red light and a searchlight, possibly linked to a camera, which throws intermittent beams of light on the "target".

Officially, the government has played down the importance of the flights. "We are neither minimising nor dramatising these events," said the Environment and Energy Minister, Ségolène Royal. "But they pose no threat to the security of nuclear sites."

Greenpeace France has accused the government of "minimising the risk" and covering up the fact that France's nuclear power stations are vulnerable to terrorist attack. "A medium-sized drone could carry an explosive charge big enough to damage the pools [of cold water] in which spent nuclear fuels are stocked," a spokesman said.

The organisation has denied any link with the flights. "Anti-nuclear campaigners tend also to be anti hi-tech," a Greenpeace source said.

Denis Baupin, a Green member of the National Assembly, said: "No one can say who is responsible for these events and no one can guarantee the good intentions of those who are controlling these objects."

France is more dependent on nuclear power than any other developed nation. Almost 80 per cent of the country's electricity comes from nuclear generators.

Source: http://www.independent.co.uk/news/world/europe/french-government-on-high-alert-after-unexplained-drone-flights-over-nuclear-power-stations-9850138.html

Iran claims successful test flight of captured US drone replica

By ASSOCIATED PRESS
The National, 10 November 2014

TEHRAN – Iran said Monday that it has successfully tested its own version of a US-made drone based on one it captured in 2011.

State TV quoted General Amir Ali Hajizadeh, head of the aerospace division of the elite Revolutionary Guard, as saying the test was carried out earlier Monday.

“As we promised earlier this year, a test flight of the Iranian version of the RQ-170 was carried out and a video will be released soon,” he said. Iran had displayed the drone earlier this year.

Iran has said it managed to reverse-engineer the RQ-170 Sentinel, seized in December 2011 after it entered Iranian airspace from neighbouring Afghanistan, and that it’s capable of launching its own production line for the unmanned aircraft.

Iran said it has captured several American drones in recent years, including a Boeing-designed ScanEagle drone — a less sophisticated aircraft — after it entered Iranian airspace over the Arabian Gulf.

Tehran said it is also reverse engineering the ScanEagle and plans to put its version into service.

Iran has accused the US of using drones to spy on the country’s nuclear programme. The US and other Western nations have long suspected Iran is secretly pursuing a nuclear weapons capability alongside its civilian programme. Tehran insists its nuclear programme is entirely focused on peaceful applications like power generation and the production of medical isotopes.

The two sides are currently struggling to reach an agreement under which Tehran would curb its nuclear activities in exchange for relief from crippling international sanctions.

Iran began a self-sufficiency military programme in 1992, under which it has produced everything from mortars and torpedoes to military tanks, jet fighters and light submarines.

Source: http://www.thenational.ae/world/iran/iran-claims-successful-test-flight-of-captured-us-drone-replica

Biodegradable drone is grown from mushrooms and covered in wasp spit

By RACHEL FELTMAN
The Washington Post, 22 November 2014

MOFFETT FIELD, CALIFORNIA – A group of college students has created an environmentally friendly drone — think veggie leather.

Led by one of NASA's synthetic biology experts, they made an unmanned aerial vehicle almost entirely out of biodegradable materials. After a crash, these little fliers would basically disappear.

Drones can be a great help in dealing with environmental issues, flying into protected wooded areas to count the surviving population of an endangered animal, or over remote coral reefs to assess their condition.

But sometimes they can turn into litter: If a drone goes down in a protected area, it might not be possible for anyone to retrieve the hunk of metal and plastic. "I have colleagues who do remote sensing in sensitive areas, and there was a UAV lost for a couple months in an area you really wouldn't want to lose one in," said Lynn Rothschild of NASA's Ames Research Center.

Rothschild serves as an adviser for a team competing in the International Genetically Engineered Machine (iGEM) competition, and the issue of downed-drone-litter seemed like a good one for her group to tackle.

"Normally I just give them free rein, but then there are 15 very bright students who all want to do 15 or 30 different things," she said. "So this year, I suggested an overall project. But they really just ran with it from there."

One of her students found a company called Ecovative Design that was growing the team's dream material: Blocks of fungal foam.

Mushrooms are made up of a structure called mycelium. It grows looking almost like a spider web when it's spread out, but it can grow to fit the confines it's placed in, eventually forming a tough chunk of foamy material. By putting mycelium into a mold filled with a tasty growing medium — like dead leaves or straw — you can create a custom-shaped mushroom block. Or in this case, a custom-shaped mushroom drone frame. A blast of heat kills the mycelium to freeze their growth.

"You end up with this great material that just leftover fungal bits," Rothschild said.

To make the frame more durable, the students created a bioplastic to coat it with.

You can make a kind of veggie leather using bacteria that create cellulose — the tough stuff that creates cell walls in plants. The bacterial cellulose is grown in a sheet and harvested, then wrapped around the mycelium frame. When it dries, it's tough and hard.

But the team's biohacking didn't stop there: They also harnessed the power of the insect world to keep their drone from dissolving in mid-air. The drone is covered in proteins cloned from paper wasp saliva, which the insects use to waterproof their nests.

For now, that's as far as the drone's biodegradability goes: It still uses a traditional rotor, battery, and controls. But other researchers around the world are already working on creating biodegradable versions of these components, Rothschild said. And her team is investigating the use of biological sensors, which would allow them to replace some of the sensors on the drone with bacteria.

"Eventually I'd say that most, if not all, of the drone could be made from biological materials," she said.

Rothschild is excited about the ways the drone could be used in research on Earth, but she has bigger plans for them, too: She's already submitted a proposal to NASA to push this tech forward for Mars missions. The lightweight, unobtrusive, home-grown nature of the robots would make them perfect for use on the red planet.

Source: https://www.washingtonpost.com/news/speaking-of-science/wp/2014/11/22/this-biodegradable-drone-is-grown-from-mushrooms-and-covered-in-wasp-spit

Cruise missile-spotting balloon rises over Maryland

By IAN DUNCAN
The Baltimore Sun, 26 December 2014

BALTIMORE – When the weather's good, a strange speck will be visible on the Baltimore skyline: a giant balloon the Army is floating high above Middle River to scan for cruise missiles. The balloon — technically an aerostat because it's tethered to the ground — is expected to launch Saturday. It's one of a pair planned for the skies over Maryland to test sensors that can detect and train fire on cruise missiles from hundreds of miles away.

They're aimed at filling what some military officials see as a gap in US air defenses. But the program has been beleaguered by shrinking budgets and opposition from privacy advocates, who warn that the balloons could be transformed into ever-present eyes in the sky, watching the movements of everyone for miles around.

Aberdeen is hosting the balloons for a three-year test. Officials stress they will not carry cameras. The Army showed off the first helium-filled balloon last week at an Aberdeen Proving Ground annex. The gleaming white, 80-yard-long inflatable hovered just above a wide patch of gravel, where it was free to swing about in the breeze.

At the briefing, Capt. Matt Villa, a planning officer with the Army, attempted to assuage concerns. "I can't stress enough there are absolutely no cameras or video equipment on board ... the system," Villa said. "Its radars cannot detect people, it does not store information ... it has no weapons on board." The military's stated intentions notwithstanding, privacy advocates say the balloons, which can float 2 miles in the air for up to a month at a time, could easily be turned into powerful spying platforms. David Rocah, an attorney with the American Civil Liberties Union of Maryland, described them as the latest example of battlefield technology being brought home for domestic use.

The Joint Land Attack Cruise Missile Defense Elevated Netted Sensor System — or JLENS — uses sophisticated radar to stare out 340 miles in every direction in search of cruise missiles and other airborne threats. The balloons work in pairs: One looks for potential threats; the other helps weapons lock onto those threats and, it is hoped, shoot them down. Data from the sensors are passed down to the ground through the tether and on to Aberdeen Proving Ground.

Maj. Gen. Glenn Bramhall, whose unit oversees the project, told reporters that cruise missiles are becoming more widely available to US enemies. Chet Nagle, a former Navy aviator and member of the lobby group the Committee on the Present Danger, said the test will leave Washington better defended against a cruise missile attack. "There is nothing else we have that will do that," Nagle said. "Most of our in-place defensive systems are designed for ICBMs," he said, referring to the much higher-flying intercontinental ballistic missiles.

In a research paper, Nagle pointed to flights by Russian Bear bombers close enough to the United States that they could reach Washington with a cruise missile. Other hostile nations could obtain a ship-mounted missile, Nagle wrote. Still, JLENS has seen its scope and funding narrowed. Defense officials originally wanted to field as many as 16 of the balloon pairs. But after costs spiked, the program was scaled back, and the pair above Maryland now look to be the only ones in operation.

If the test proves successful, Bramhall said, the program could be expanded. But the House of Representatives attempted to slash by half the $54 million proposed budget for JLENS in this year's National Defense Authorization Act. The legislation signed by President Barack Obama this month restored most of the money, but Bramhall's team will have to make do with about $12 million less than it asked for.

> *Source: http://www.baltimoresun.com/news/maryland/harford/aberdeen-havre-de-grace/bs-md-surveillance-blimps-20141226-story.html*

Serbia's Nemanja Gudelj, left, holds a flag as he and Albania's Bekim Balaj, centre, clash during the Euro 2016 qualifying match between Serbia and Albania. ©AP/Darko Vojinovic

A biodegradable drone made from fungus. © Stanford–Brown–Spelman iGEM team 2014

اسرائیل باید از صحنه روزگار
هیچ غلطی نمی تواند بکند
out of the scene
American Can Do

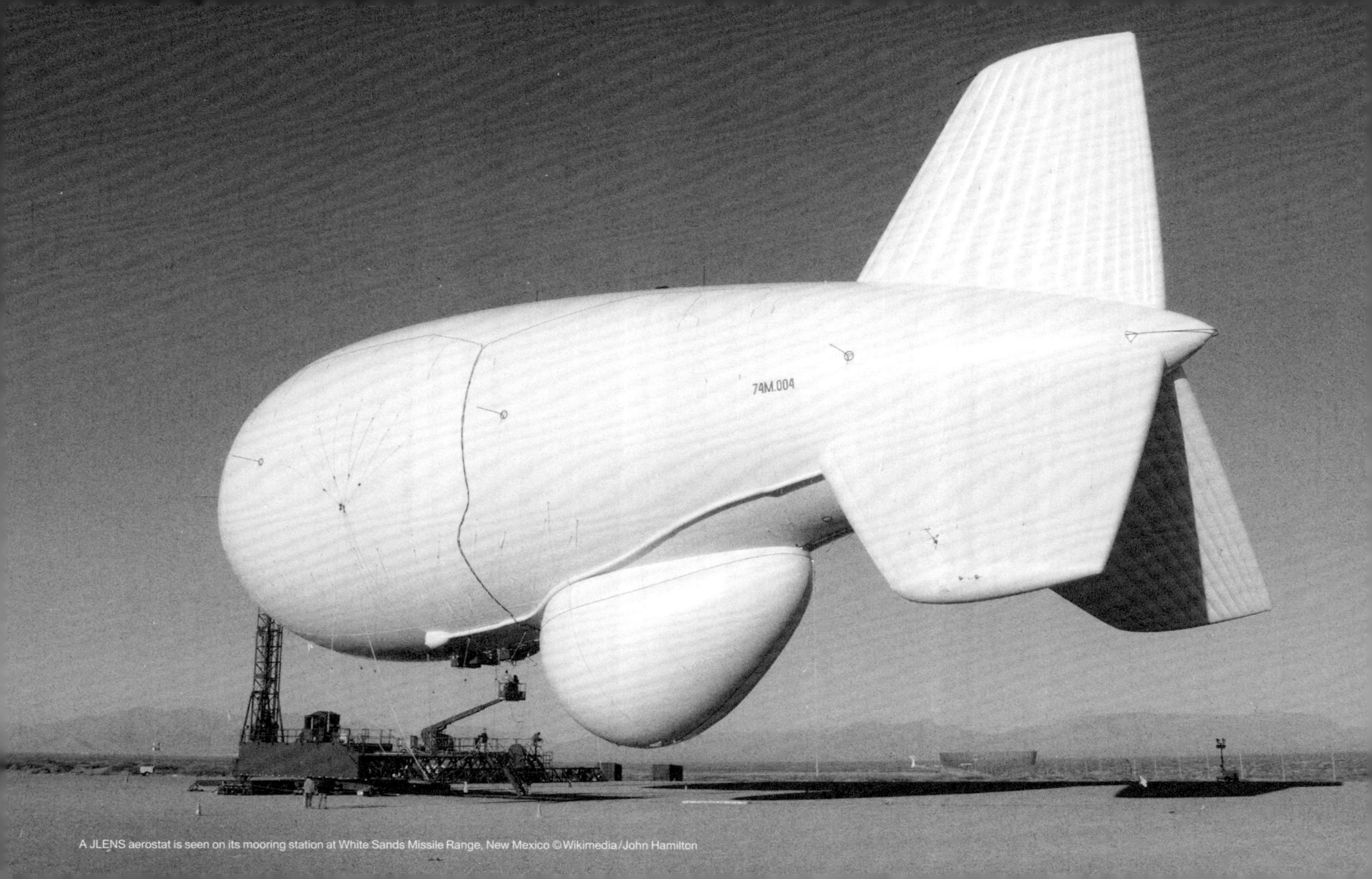

A JLENS aerostat is seen on its mooring station at White Sands Missile Range, New Mexico ©Wikimedia/John Hamilton

Drone that crashed at White House was quadcopter

By ZEKE J. MILLER
Time Magazine, 26 January 2015

WASHINGTON – A drone that crashed on the White House grounds Monday, causing a brief lockdown, was a two-foot wide remote-controlled quadcopter that is sold in stores, officials said.

According to a Secret Service spokesman, a uniformed division officer stationed on the South Grounds of the complex 'heard and observed' the device flying at a low altitude, before it crashed on the southeast side of the 18-acre secure zone around the executive mansion shortly after 3 a.m. Monday. The incident triggered a lockdown of the White House and nearby buildings, as officials scrambled to study the device and ensure it did not pose a threat.

According to Secret Service spokesman Brian Leary, an individual called the agency Monday morning after seeing news reports of the crash to report that they had been in control of the quadcopter. "The individual has been interviewed by Secret Service agents and been fully cooperative," he said. "Initial indications are that this incident occurred as a result of recreational use of the device. This investigation continues as the Secret Service conducts corroborative interviews, forensic examinations and reviews all other investigative leads."

President Barack Obama and First Lady Michelle Obama were in India when the incident occurred. It is not clear whether other family members were present.

White House Press Secretary Josh Earnest confirmed that a device had been recovered. "There is a device that has been recovered by the Secret Service at the White House," he said in a press briefing in New Delhi early Monday. "Early indications are that it does not pose any sort of ongoing threat right now to anybody at the White House."

The crash follows several high-profile security breaches at the White House that have shaken the Secret Service, including an incident last year when a disturbed man armed with a knife jumped a fence and managed to enter the mansion before being apprehended by officers. Obama subsequently asked the agency's director to step aside, and her interim replacement has taken steps to reform its top leadership.

Under longstanding Federal Aviation Administration guidelines, no unmanned aerial system may be flown in the 12-13 mile area around Washington Reagan National Airport, which includes airspace over the White House, Pentagon, Naval Observatory and the Central Intelligence Agency.

The proliferation of small drones is posing new challenges not just at sensitive government facilities, but around the country, as cheap systems equipped with cameras pose new privacy concerns and reports of close encounters with private and commercial aircraft rise.

The Secret Service released a photo of the device Monday afternoon, identifying it as a member of the popular DJI Phantom line of quadcopters which retail for several hundred dollars online. It was not immediately clear whether the device was equipped with a camera, or whether it was recording during its flight.

Source: http://time.com/3682307/white-house-drone-crash/

Dozen more drone sightings reported over Paris

By AGENCE FRANCE-PRESSE
The Guardian, 4 March 2015

Paris – About a dozen more drone sightings over sensitive areas of the French capital — the latest in a series of unexplained overflights — have been reported to Paris police by members of the public.

The authorities are investigating whether the sightings were really drone flights, said a police source who declined to be named.

No arrests have been made in connection with the flights, which reportedly took place overnight on Tuesday near the river Seine, Place de la Concorde, the Invalides military museum and around the Paris ring road.

French authorities are still trying to find a motive for the reported flights. The interior minister, Bernard Cazeneuve, announced on Tuesday there had been about 60 incidents over nuclear plants or the French capital since October.

On Tuesday, a court fined a British journalist €1,000 ((about £725) for flying a drone over central Paris and confiscated his machine.

Tristan Redman, 34, from the Qatar-based network Al-Jazeera, was arrested last month in the Bois de Boulogne park on the western edge of the city with two other journalists. The two others were released and Redman pleaded guilty to flying a drone over the capital, which is illegal under French law.

Their arrest followed two nights of unexplained drone sightings over the French capital; a source with knowledge of the case said the journalists were not involved in the earlier incidents. Those mysterious sightings were made near the US embassy, the Eiffel tower and several major roads leading in and out of the French capital, police said.

Police have been unable to catch any of the operators of the night-time flights and it is unclear whether they were the work of pranksters, tourists or were something malicious.

France remains on high alert after jihadi attacks in and around the French capital in January left 17 people dead.

Source: https://www.theguardian.com/world/2015/mar/04/drone-sightings-paris-seine-concorde-invalides

Drones beaming web access are in the stars for Facebook

By QUENTIN HARDY & VINDU GOEL
The New York Times, 26 March 2015

MENLO PARK, CALIFORNIA – In classical mythology, Aquila is the eagle carrying Jupiter's thunderbolts skyward. At Facebook, it is the code name for a high-flying drone, indicative of the social networking company's lofty ambitions.

The V-shaped unmanned vehicle, which has about the wingspan of a Boeing 767 but weighs less than a small car, is the centerpiece of Facebook's plans to connect with the five billion or so people it has yet to reach.

Taking to the skies to beam Internet access down from solar-powered drones may seem like a stretch for a tech company that sells ads to make money. The business model at Facebook, which has 1.4 billion users, has more in common with NBC than Boeing.

But in a high-stakes competition for domination of the Internet, in which Google wields high-altitude balloons and high-speed fiber networks and Amazon has experimental delivery drones and colossal data centers, Facebook is under pressure to show that it, too, can pursue projects that are more speculative than product.

One of those offbeat ideas, or so the thinking goes, could turn out to be a winner.

"The Amazons, Googles and Facebooks are exploring completely new things that will change the way we live," said Ed Lazowska, who holds the Bill and Melinda Gates Chair in Computer Science and Engineering at the University of Washington. "There are other companies out there like Hewlett-Packard and IBM, but they aren't doing the really huge things anymore."

At a conference in San Francisco on Wednesday, Mark Zuckerberg, Facebook's chief executive, explained how Facebook was opening the code to Messenger, its mobile messaging app, so other companies can build right on top of what it's already doing. The move is a key component of its efforts to improve its already strong position among mobile computing devices.

That is Step 1 in a series — including drones, goggles that plug into virtual reality, and artificial intelligence — Facebook plans to work on in the coming years to broaden its influence.

At the conference, the company showed off nearly 50 apps for Messenger, including one that turns text messages into songs and another that allows a user to search for and send an animated GIF to express a mood.

"Until now, we have focused on improving Messenger by building all of these features ourselves," Mr. Zuckerberg told developers.

By opening the platform, he said, Facebook wants to tap the creativity of other app makers. "We're really excited about what you all are able to do with these fast and simple tools we are rolling out," he said.

Facebook said it would soon begin testing a system to allow businesses to use Messenger to offer personalized service to customers after a purchase is made. For example, a person might be able to use Messenger to change the color of a shirt in an order or see where the package is in the UPS delivery system.

Facebook's initial partners are the clothing retailers Zulily and Everlane, and the company hopes to lure other types of businesses, like airlines and cable companies, that frequently face customer service issues.

"We all have to agree that it's pretty painful to interact with businesses right now," David Marcus, Facebook's head of messaging products, said in an interview. "We're bringing back the conversational nature of commerce."

Facebook also demonstrated a new type of post in its news feed that allows users to click and explore a scene in 360 degrees.

Next up: Connecting with the rest of the world, which will be the main topic for Day 2 of the company's conference.

Facebook's drone team, which came to the company through the acquisition last year of the drone maker Ascenta, say they believe their solar-powered craft can eventually be aloft up to three months at a time, beaming high-speed data from 60,000 to 90,000 feet to some of the world's remotest regions via laser. Test flights are to begin this summer, though full commercial deployment may take years.

"We want to serve every person in the world" with high-speed Internet signals, said Yael Maguire, head of Facebook's Connectivity Lab. The dream — assuming regulators around the planet go along with it — is a fleet as big as 1,000 drones connecting people to the Internet. And where it is too remote even for the drones, satellites would do the trick.

"Can we reach a point where everyone on the planet gets the same message at once?" Mr. Maguire said. "I'm looking forward to that day."

That day could be a long way off, Facebook executives admit. While they don't know what a final version of Aquila will cost, they appear comfortable with spending billions to realize that vision. Others are not so sure.

"Like Google, Facebook gets a pass, because they've defied critics and are run by visionary leaders looking out five or 10 years," said Scott Kessler, an analyst with S.&P. Capital IQ. "Still, people would like to know what this costs, and if it makes money."

Facebook is also courting partners with the manner in which it builds new kinds of computers at its data centers. The idea is that if it is open about what it does and even allows other people to use that information and experiment with their own prototypes, then more companies will adopt that technology and align themselves with Facebook.

Hewlett-Packard, for example, recently said it would sell computer servers using Facebook designs, which should lower Facebook's own costs when it buys those servers from HP, or buys semiconductors from the suppliers it has in common with the company.

"You go into our data centers and it's hard to find anything that wasn't built by us," said Michael Schroepfer, Facebook's vice president of engineering. "We want tons of other companies out there that service the entire planet. We will service the core needs better than anyone else, for free."

Facebook's effort in artificial intelligence is called deep learning, for the number of levels at which it critically analyzes information. By figuring out context, Facebook better knows why people anywhere are looking at something, and what else it can do to keep them engaged.

Twitter and even Google collaborate with Facebook and others on open-source A.I. software called Torch. Facebook says it likes working with competitors on that technology because, in the end, they can help improve it.

"There is nothing that any single company is going to solve by itself," said Yann LeCun, Facebook's A.I. research director and a New York University professor.

For the long term, Mr. Zuckerberg hopes Facebook's A.I. will translate languages on the fly, know strangers you might meet and, of course, bring you the highest-value ads.

Because, in the end, it's still about getting you to look at more ads.

"The fundamental thing about advertising is people paying to get a message in front of you," Mr. Schroepfer said. "That won't go away in my life, though the form may change."

Source: http://www.nytimes.com/2015/03/26/technology/drones-beaming-web-access-are-in-the-stars-for-facebook.html

Airmail via drones is vexing for prisons

By MICHAEL S. SCHMIDT
(KITTY BENNETT contributed research)
The New York Times, 22 April 2015

BISHOPVILLE, SOUTH CAROLINA – During the graveyard shift at 1:44 a.m., security cameras at the prison here picked up the blinking lights of an unidentified flying object approaching the facility's fence.

A corrections officer was dispatched to investigate, but by the time she got there, all she could see was a man running away into the dense forest that surrounds the prison.

It was not until dawn that officers found a package that included a cellphone, tobacco and marijuana tangled in the power lines outside the prison and a small drone that had crashed in the bushes nearby. In the woods, investigators located a makeshift campground, the remote control device used to fly the drone, a bottle of grape-flavored Gatorade and drugs.

"It was a delivery system," said Bryan P. Stirling, the director of the South Carolina Department of Corrections, explaining how the drone's operators had planned to send the contraband into the prison, the Lee Correctional Institution. "They were sending in smaller amounts in repeated trips. They would put it on there, they would deliver it, someone inside would get it somehow, and they would send it back out and send more in."

It is the high-tech version of smuggling a file into a prison in a birthday cake, and it underscores the headache that drones are now creating for law enforcement and national security officials, who acknowledge that they have few, if any, ways of stopping them.

Drones flying over prison walls may not be the chief concern of corrections officials. But they say that some would-be smugglers are experi-menting with the technique as an alternative to established methods like paying off officers, hiding contraband in incoming laundry and throwing packages disguised as rocks over fences into recreational yards.

The authorities have detected at least three similar attempts at corrections facilities in the United States in the past two years. In the same period, there were also at least four reported attempts abroad, in Ireland, Britain, Australia and Canada.

In January, guards found a drone with blue and red flashing lights on the ground inside a recreational yard at a prison in Bennettsville, S.C., according to investigative reports. On that drone were 55 grams of synthetic marijuana and a cellphone charger, the reports said.

Law enforcement officials say they have no way of knowing how many attempts have been successful, but the warden of the Lee Correctional Institute, Cecilia Reynolds, said that in recent weeks her officers found 17 phones in one inmate's cell. She said she suspected that the phones continue to come in on drones.

"We've got to do something about this — these cellphones are killing us," she said.

Smartphones are so desirable to inmates because unlike pay phones at prisons, they are not recorded or monitored.

The phones also allow them to watch pornography and communicate surreptitiously with fellow prisoners. The phones are essential for coordinating with smugglers using drones, because the prisoners need to know where to find the deliveries in the yard. The prisoners can then use the phones to quickly pay their suppliers.

Ever since a hobbyist drone crashed on the South Lawn of the White House in January, there has been an increased push by law enforcement agencies, companies and privacy watchdogs to come up with ways to restrict the flight of drones. That task has been particularly difficult because the laws and regulations that govern drones are outdated and confusing.

"It's a new area and we don't know the full extent of drones' capabilities — both good and bad — and because of that we don't know what the gaps are in the laws and what we need to do to clarify them," said Lisa Ellman, who helps lead the unmanned aircraft systems practice group at the law firm McKenna, Long & Aldridge in Washington.

Nets cannot be built over every sensitive location in the United States to keep drones out. So, for now, the best way to control the problem appears to be 'geofencing' software that prevents the drones from flying over a specific location.

DJI, the company that manufactured the drone that crashed at the White House, announced in March that its new geofencing software would make its devices inoperable within roughly 16 miles of the White House. The company said it was working to create similar no-fly zones for 'sensitive institutions and national borders.'

An organization called No Fly Zone has introduced a website where individuals, business owners and others who do not want drones overhead can enter their addresses into a database. Those addresses will be provided to drone manufacturers who have agreed to program their devices not to fly over those locations.

For several years, officials at the prison here have tried to deter people from throwing objects over its fences by installing dozens of beehives in the woods around it. In front of the hives are signs saying: "Danger! Bees! Keep Out!"

But in response to the drone incidents, the authorities in South Carolina have taken several measures that they say will make it harder for drones to fly over their prisons. For example, they said that with the backing of Gov. Nikki R. Haley they had built hulking new towers that enable officers to look outward into the areas around the state's correction facilities.

"It's almost like we need an Iron Dome like Israel has to stop it," Mr. Stirling said, referring to the multibillion-dollar rocket interceptor system that was used to protect Israel during its most recent conflict with Hamas. "But they have a robust defense budget."

After the drone incident here last April, investigators found a receipt from a convenience store at the campground where the smugglers were hiding. The investigators went to the store and obtained images of the men from surveillance cameras.

Two months later, they arrested one of the men, Brenton Lee Doyle, who is now serving his sentence at another state prison. That arrest led the authorities to a second man, who was arrested last week in Tennessee.

Mr. Stirling said he never expected to have to learn so much about drones when he became the head of the state's prison system in 2013.

"When I started in this job it was all very futuristic — Amazon wasn't even talking that much at that point of using them to make deliveries," he said. "Now it's something we're having to devote extensive resources to."

"We put up higher fences to stop people from throwing things over them," he said. "Now they're just flying over them."

Source: http://www.nytimes.com/2015/04/23/us/drones-smuggle-contraband-over-prison-walls.html

Drone 'containing radiation' lands on roof of Japanese PM's office

By THE ASSOCIATED PRESS
The Guardian, 22 April 2015

TOKYO – Japanese authorities have launched an investigation after a small drone reportedly containing traces of radiation was found on the roof of the prime minister's office, sparking concerns about drones and their possible use for terrorist attacks.

No injuries or damage was reported from the incident on Wednesday. The prime minister, Shinzo Abe, who is at present in Indonesia, works at the building during the day and commutes from his own private home roughly 15 minutes away.

Police said it was not immediately known who was responsible for the drone.

The significance was not immediately clear, but a Japanese court on Wednesday approved the restart of a nuclear power station, rejecting nuclear power safety worries in the wake of the 2011 Fukushima radiation disaster.

The chief cabinet secretary, Yoshihide Suga, said the incident was a wakeup call to the potential dangers of drones, including possible terror attacks when Japan hosts the 2020 Olympic Games in Tokyo.

It was not clear when the drone landed. It was found on Wednesday by an official who was reportedly taking new employees on a tour of the prime minister's office in central Tokyo.

Japanese media reported that investigators detected small traces of radiation from the drone, presumably not levels harmful to humans. TV footage showed several uniformed policemen without hazmat suits carrying a blue plastic box containing the drone for further examination.

Video from public television broadcaster NHK earlier showed dozens of police officers and officials around the drone, which was covered by a blue tarpaulin.

The drone was about 50cm in diameter and had four propellers, carrying a small camera and a plastic bottle with unidentified contents inside, according to Tokyo's Metropolitan police department. The drone was also decorated with a symbol that warns of radioactive material, according to NHK.

Small drones are becoming increasingly popular in Japan and are often used for performances, aerial filming and other purposes, but have been raising safety concerns.

In the United States in January, a wayward drone flown by an off-duty intelligence employee crashed on the White House grounds, raising questions over how commercial and consumer drones can be used safely in the US.

Japanese aviation laws have no restrictions for flying unmanned equipment at or below 250 metres above ground except for flight routes.

Source: https://www.theguardian.com/world/2015/apr/22/drone-with-radiation-sign-lands-on-roof-of-japanese-prime-ministers-office

Warren Weinstein and the long drone war

By STEVE COLL
The New York Times, 23 April 2015

WASHINGTON – "As President and Commander-in-Chief, I take full responsibility for all our counterterrorism operations," President Obama said in a press conference at the White House on Thursday. Obama was announcing the news that C.I.A. drone strikes in Pakistan last January killed two hostages being held by Al Qaeda — Warren Weinstein, an American, and Giovanni Lo Porto, an

Italian — along with two American-born Al Qaeda members. Obama added, "I profoundly regret what happened."

That is what a President ought to say at such a moment. Obama owns the decision to prioritize lethal drone strikes over other counterterrorism strategies. He owns the record of civilian and other collateral deaths that those strikes have created — a record still frustratingly shrouded in secrecy. But the President's mea culpa masks an important and timely question: How exactly did this mistake at the C.I.A.'s Counterterrorism Center happen, and who will be held accountable for it? By putting the President out front and withholding all but the broadest details about the failed operation, the Obama Administration apparently hopes to evade that question.

There is circumstantial evidence that the Administration has been preparing for this announcement for some weeks. On March 25th, news organizations reported that the powerful longtime leader of the C.I.A.'s Counterterrorism Center, a dyspeptic Muslim convert whose code name is Roger, was being replaced. The Center carries out the great majority of covert drone strikes; the Pentagon carries out a smaller number. Roger, who took full charge of drone operations about nine years ago, had become perhaps the most powerful figure inside the C.I.A. since the days of James Angleton, who, as the agency's chief of counterintelligence from 1954 to 1975, was charged with identifying Soviet moles.

When the news of Roger's ouster emerged — he was to be replaced with a former Kabul station chief and operations officer in South Asia — anonymous officials characterized him as a casualty of larger bureaucratic reorganizations. Was that all it was? After word of Weinstein and Lo Porto's killing first materialized, and then was investigated, and then was confirmed, who documented and evaluated the causes of the botched drone strike? What form did the investigation take? What did the evidence show? What were the findings and recommendations?

The idea that detailed answers to such questions should lie beyond public or congressional scrutiny simply because drone operations are secret does not pass the laugh test. When covert drone operations and sensitive decision-making protocols are depicted on 'Homeland,' with colorful details drawn from the coöperative explanations of former C.I.A. officials, they no longer qualify as state secrets, and certainly not ones important enough to trump the democratic imperatives of transparency and accountability.

The C.I.A. and the Obama Administration have sold their policy on the proposition that drone strikes — and the intensive, sometimes days-long aerial surveillance that typically precedes them — are a highly discriminating form of warfare. In a hostile and inaccessible environment like the western tribal areas of Pakistan, they claim, a drone is preferable to the alternative, which might include a conventional bombing attack, a helicopter raid, or a more passive containment strategy in collaboration with (sometimes compromised) local security forces. A Nobel Peace Prize laureate's embrace of drones, partly on humanitarian grounds, is sure to increase their legitimacy as instruments of war in the future. But how can Obama's choice be squared with the accumulating record of mistakes?

The question must be contextualized. Obama has actually presided over two drone policies — one during his first term, the other since 2013. During his first term, Obama oversaw an undeclared air war in the Federally Administered Tribal Areas of Pakistan, a drone campaign designed not only to put pressure on Al Qaeda but also to support the Administration's 'surge' of ground troops in Afghanistan. The pace of strikes in Pakistan reached two per week in 2010. The campaign produced substantial but unknown numbers of civilian casualties; the Administration's unpublished count of civilian deaths, heavily influenced by the C.I.A.'s grading of its own homework, runs only into the dozens, but the counts of independent researchers, relying on local press reporting and field interviews, runs into the hundreds. The Administration has refused to make its accounting of civilian deaths during these violent years available for any sort of scrutiny or redress.

In May, 2013, under pressure from human-rights groups and European governments, Obama announced a rewrite of the rules for lethal drone attacks. The Administration disclosed that it would now only allow strikes to be launched if there was a 'near certainty' of zero civilian casualties. After that, the number of strikes fell substantially. Now, however, Obama must own up to the fact that the changes he ordered failed to prevent a tragic, consequential error. The best way for the President to accept responsibility would be for him to order greater transparency about what happened in January and why, and, more broadly, to accept that secret violence beyond public accountability, no matter its justification, cannot play such a central role in any democracy's 'long war' strategy.

One aspect of this particular mistake in Waziristan colors the news with a distinctive sadness. Warren Weinstein was the forgotten man of the war against Al Qaeda. He was an Urdu-speaking aid worker on contract with U.S.A.I.D., a man past retirement age, who was kidnapped from his home, in Lahore, in the summer of 2011, days before he was supposed to return to the States. The kidnapping occurred three months after Navy seals raided a compound in Abbottabad, Pakistan, and killed Osama bin Laden. Yet, despite efforts that the Obama Administration described on Thursday as extensive, no seals ever located or attempted to rescue Weinstein, who was seventy-three years old when he died.

Nor did the White House negotiate his release. Last May, after long talks with the Taliban, US Special Forces flew into

Waziristan to accept custody of Bowe Bergdahl, an Army soldier who had wandered off his base, on the Afghan border with Pakistan, and been captured by fighters with the Taliban-affiliated Haqqani network. In exchange for Bergdahl's release, the Obama Administration released four Taliban prisoners held at Guantánamo. Weinstein did not figure into the deal and was left behind. Judging by the videos that his captors released, he was ill and deeply demoralized.

Of course, Al Qaeda, not the Obama Administration, is responsible for Weinstein's miserable fate. Still, the fact that Weinstein's own government accidentally killed him — during his fourth year in captivity, and without a rescue ever being attempted — is a disturbing coda to the short history of drone warfare. It reminds us that the problem with drones is not just that their operators sometimes make mistakes. It is that the heavy reliance — in time, dollars, and bureaucratic priorities — on a technological panacea for the problem of terrorism can cause a government to lose sight of the people on the ground.

Source: https://www.newyorker.com/news/news-desk/warren-weinstein-obama-drone-war

Protective dad uses drone to follow his daughter as she walks to school

By KARA O'NEILL
The Mirror, 24 April 2015

KNOXVILLE, TENNESSEE – A protective father has come up with a unique way to make sure he can always keep an eye on his daughter — by following her with a drone.

Chris Early, who owns a video production company in Knoxville, Tennessee, bought the device to create advertisement campaigns but has since found it comes in useful for his own surveillance purposes.

The concerned dad admitted he was uneasy when his eight-year-old daughter asked if she could make the trip to school alone, but it didn't take him long to figure out a sneaky way to check up on his little girl.

Using the drone to hover in the air above his daughter, Chris was able to put his mind at rest and make sure she got to school safely.

In the clip, the little girl can be seen sticking to the grass verge as she walks along a residential street to her school.

Chris' daughter didn't realised her dad had followed her to school but has said she thinks her father's idea was a cool one.

Chris told ABC News: "It was just kind of a thing, just to kinda keep an eye on, just to make sure she was looking both ways." "Let her know that daddy is always watching."

Since the footage emerged online, Chris has received a flurry of comments criticising his parenting style for being too 'invasive'.

Responding to the backlash, Chris explained: "The drone flight following my daughter [to] school was a one-time thing done for fun."

The tech enthusiast added: "It was never meant to set off a debate about how to parent in the 21st century."

"We actually live very close to her school [and] I can just about see the school from my front porch without a drone."

Source: https://www.mirror.co.uk/news/technology-science/technology/protective-dad-uses-drone-follow-5572475

The age of drone vandalism begins with an epic NYC tag

By ARTHUR HOLLAND MICHEL
wired.com, 30 April 2015

NEW YORK – In the early hours of Wednesday morning, the age of robotic graffiti was born. KATSU, a well-known graffiti artist and vandal, used a hacked Phantom drone to paint a giant red scribble across Kendall Jenner's face on one of New York City's largest and most viewed billboards. By all accounts, it is the first time that a drone has been deployed for a major act of public vandalism.

In April last year, KATSU made headlines when he demonstrated that he had figured out how to attach a spray can to an off-the-shelf DJI Phantom drone. At the time, he was only using the drone to paint canvasses for white-wall galleries. But he assured the world that soon he would take his mad invention out into the streets and create enormous tags in places that were previously inaccessible to even the most daring and acrobatic taggers. Now, he appears to have made good on his promise in grand fashion.

"It turned out surprisingly well," said KATSU, whose previous stunts include using a hacked fire-extinguisher to vandalise L.A. MOCA. "It's exciting to see its first potential use as a device for vandalism," he added, cheerfully.

The Calvin Klein billboard, one of New York City's largest, sits at the busy intersection of Houston St and Lafayette St. The graffiti drone's potential for troublemaking on an unprecedentedly grand scale is obvious. The billboard, which was previously graced by a topless, (perhaps) digitally-enhanced Justin Bieber, is absolutely gigantic, about six stories tall. It would have been almost impossible to tag Jenner's face using the traditional methods. One could rappel off the top of the building or use a cherry picker, but neither option is exactly safe, or subtle, or quick enough that one could do it without cops on regular patrol spotting it. With the drone, by contrast, it took less than a minute. Still, the artist admitted, "It was a bit tense." (Needless to say, the stunt was extremely illegal).

As the domestic drone industry grows feverishly, and multicopters like DJI's Phantom become cheaper and more powerful, artists have been eager to experiment with the technology. It was only a matter of time, then, that people would figure out that the drone has enormous potential for subversive acts on the streets, where defying the laws of gravity is the whole point. Given the enduring privacy, safety, and legal concerns around the technology, conceptually it makes a certain amount of sense that it would find uses at the peripheries of what most people (let alone the law) would consider acceptable. KATSU's scribble high above SoHo might not look like much, but it represents the potential that drones have to transform graffiti forever.

Still, police departments across the country probably don't need to start panicking quite yet. This is, after all, graffiti drone 1.0. KATSU said that it can be temperamental and unpredictable, especially when it shifts perpendicular to the surface that its painting. The controls can be twitchy. "Seventy percent of the concentration is in maintaining this equilibrium with the two dimensional surface while you are painting," he explained. We have a ways to go until drones are capable of autonomously blasting tags while their artist masters relax at home.

But that is the plan, and KATSU's stunt this week was proof of concept. He is also gearing up to release a new, more user-friendly version of the graffiti drone "very soon." While he refused to give me too many details, he did say that it would have some element of computer vision to help with stability.

Still, even graffiti drone 1.0 is something to be reckoned with. It has made what was up until yesterday an impossible tag look easy. KATSU himself seems to have been caught a little off guard by how powerful the drone has proven to be. "It's a bit frightening."

Source: http://www.wired.com/2015/04/age-drone-vandalism-begins-epic-nyc-tag/#slide-1

Re-Planting a forest, one drone at a time

By JAMES DALY
wired.com, 10 May 2015

Oxford, UK – That funny little buzz you hear in the forest may not just be the hum of summer insects. In the near future it could be a small fleet of drones, coming to replant and restore forests that have been stripped of trees by industrial-scale deforestation.

It's all part of an ambitious plan by BioCarbon Engineering, a UK-based startup on a global mission to battle widespread clear-cutting, which strips more than 26 billion trees off the planet each year. CEO Lauren Fletcher, who spent 20 years as an engineer with NASA, says the only way to fight industrial-scale deforestation is with industrial-scale reforestation. Their idea: plant 1 billion trees a year. The first targets are in South Africa and the Amazonian jungles, both of which have suffered from widespread forest eradication.

BioCarbon's reforestation scheme is simple and efficient. Here's a quick look at how it plans to deploy its drone fleet:

Do a 3-D aerial survey.
First, drones are sent to fly over a potential planting zone, snapping photos that create 3-D maps of the area to be reforested. The number of drones will vary depending up on the size of the seeding.

Create a seeding plan.
Once all that terrain data has been analyzed, it then generates a seeding pattern that best suits the terrain.

Load the seed pods.
The drones, which are equipped with guidance and control software, carry pressurized canisters of seed pods with germinated seeds immersed in a nutrient-rich gel.

Hover and plant.
Flying at a height of 1 or 2 meters, the drones follow the planting patterns, firing the biodegradable seed pods down to the ground. The pods break open upon impact, allowing the germinated seed a chance to take root.

Monitor growth.
After planting, the drones do low-level flights to assess the health of the sprouts and saplings.

Such 'precision forestry,' as BioCarbon calls it, is extremely efficient. A farmer might hand plant as many as 3,000 seeds a day; Fletcher says his drones can drop up to 36,000 seed pods daily, often in areas where a human can't reach. Working with local ecologists, BioCarbon will use the drones to spread a variety of tree species, as well as microorganisms and fungi designed to improve the soil quality. "The central focus is ecosystem restoration," Fletcher says.

Fletcher sees the drones, which are commercially built by companies like Vulcan UAV and then modified, as an important way for existing reforesting organizations to expand their approach. "There are some times when planting by hand is absolutely the right approach," he says. "But, in other instances, the drones can be a very effective tool for the right location at the right time."

The full implementation will have a team of two operators running seven or eight drones simultaneously. Planting at about 10 pods per minute will equate to roughly 36,000 trees per day for each team. With 100 two-member teams — BioCarbon's goal in the next 5-7 years — it expects to plant 1 billion trees a year over roughly 500,000 hectares.

BioCarbon has funding from the Skoll Foundation and recently was featured in a Drones for Good competition in the United Arab Emirates. It plans to begin field testing by September, and its efforts can't come fast enough. A new study published by the American Geophysical Union (AGU) says that the rate of tropical deforestation has risen by 62 percent between the 1990s and 2000s. One reason is that tropical deforestation has become more devastatingly efficient, notes geographer Douglas Morton of NASA's Goddard Space Flight Center. "In the 60s, it was axes; in the 70s, chainsaws; and in the 2000s, it was tractors."

> *Source: http://www.wired.com/brandlab/2015/07/re-planting-forest-one-drone-time/*

Enrique Iglesias injures hand grabbing drone during concert

By DANIEL KREPS
rollingstone.com, 31 May 2015

TIJUANA – Enrique Iglesias injured his hand during a concert in Tijuana, Mexico Saturday after the singer attempted to grab a drone that was filming the show. Iglesias' fingers were sliced and bloodied following the incident, but the bandaged singer continued on with the rest of the performance. After the show, Iglesias hopped on a plane to Los Angeles to see a specialist for his injured hand, *USA Today* reports.

"During the show a drone is used to get crowd shots and some nights Enrique grabs the drone to give the audience a Point of View shot. Something went wrong and he had an accident," Iglesias' representative said in a statement. "He was semi-treated by crew on the side of the stage to try and stop the bleeding. He was advised to stop the show. He decided to go on and continued playing for 30 minutes while the bleeding continued throughout the show."

In the video of the incident, Iglesias uses the drone's camera to essentially selfie himself for the 12,000 visitors at Tijuana's Plaza de Toros de Playas. However, while harnessing the drone, Iglesias accidentally slid his fingers into the rotor's path, resulting in the injury. The singer then tossed the drone to the floor and bloodied up his white T-shirt.

Iglesias will have plenty of time to recover as he's next scheduled to perform in Mexico City on July 3rd.

> *Source: http://www.rollingstone.com/music/videos/enrique-iglesias-injures-hand-grabbing-drone-during-concert-20150531*

China uses drones to catch students cheating in exams

By DANIELLE DEMETRIOU
The Telegraph, 5 June 2015

BEIJING – Officials in China are taking extreme measures to prevent students from cheating in famously tough university entrance examinations — by using drones.

The high-tech method is being used by education authorities in Luoyang, central China, in a bid to curb students from cheating, according to Chinese media reports.

Silent-flying drones will be used to monitor students during university entrance exams known as "gaokao", which are taken by more than nine million teenagers every year.

The devices, which can hover in the air for up to half an hour, will use 360 degree rotations to scan examination halls and pinpoint the exact location of suspicious radio signals.

From heights of up to 1,640 feet, the drones will be able to hone in radio signals created by students who are using hidden earpieces to obtain the answers to exam questions, according to the People's Daily Online.

The two-day examinations are famously gruelling for students, with university admissions and subsequently their future careers paths almost entirely dependent on the results.

The high-level pressure is fuelled further by a shortfall of places, with only 6.5 million admissions accepted, resulting in one in three students unsuccessful in their bid to continue studying.

Students have reportedly cheated in the past by using a special pen that takes pictures of questions and transmits them to an outside accomplice who in turn provides answers via a secret earphone.

The use of drones is the latest weapon employed by the authorities in a long standing battle against cheating among not only students but also teachers and parents.

The tough examinations reportedly prompted officials at one secondary school in Hebei province to install cage-lie "anti-suicide" barriers earlier this year to prevent students jumping to their deaths.

Two years ago, education chiefs in Jilin province took the unusual step of banning bras with metal clips in a bid to deter students from using increasingly sophisticated cheating technology.

News of the drones came just days after the Education Ministry reportedly issued a statement announcing a nationwide crackdown on cheating, according to the WantChinaTimes.com.

Last year, the Education Ministry also announced that it was planning to change the structure of university admissions exams in a bid to reduce the intense pressure the exams place annually on students and their families.

> *Source: http://www.telegraph.co.uk/news/worldnews/asia/china/11654236/China-uses-drones-to-catch-students-cheating-in-exams.html*

As stress drives off drone operators, Air Force must cut flights

By CHRISTOPHER DREW & DAVE PHILIPPS
The New York Times, 16 June 2015

Creech Air Force Base, Nevada – After a decade of waging long-distance war through their video screens, America's drone operators are burning out, and the Air Force is being forced to cut back on the flights even as military and intelligence officials are demanding more of them over intensifying combat zones in Iraq, Syria and Yemen.

The Air Force plans to trim the flights by the armed surveillance drones to 60 a day by October from a recent peak of 65 as it deals with the first serious exodus of the crew members who helped usher in the era of war by remote control.

Air Force officials said that this year they would lose more drone pilots, who are worn down by the unique stresses of their work, than they can train.

"We're at an inflection point right now," said Col. James Cluff, the commander of the Air Force's 432nd Wing, which runs the drone operations from this desert outpost about 45 miles northwest of Las Vegas.

The cut in flights is an abrupt shift for the Air Force. Drone missions increased tenfold in the past decade, relentlessly pushing the operators in an effort to meet the insatiable demand for streaming video of insurgent activities in Iraq, Afghanistan and other war zones, including Somalia, Libya and now Syria.

The reduction could also create problems for the C.I.A., which has used Air Force pilots to conduct drone missile attacks on terrorism suspects in Pakistan and Yemen, government officials said. And the slowdown comes just as military advances by the Islamic State have placed a new premium on aerial surveillance and counterattacks.

Some top Pentagon officials had hoped to continue increasing the number of daily drone flights to more than 70. But Defense Secretary Ashton B. Carter recently signed off on the cuts after it became apparent that the system was at the breaking point, Air Force officials said.

The biggest problem is that a significant number of the 1,200 pilots are completing their obligation to the Air Force and are opting to leave. In a recent interview, Colonel Cluff said that many feel "undermanned and overworked," sapped by alternating day and night shifts with little chance for academic breaks or promotion.

At the same time, a training program is producing only about half of the new pilots that the service needs because the Air Force had to reassign instructors to the flight line to expand the number of flights over the past few years.

Colonel Cluff said top Pentagon officials thought last year that the Air Force could safely reduce the number of daily flights as military operations in Afghanistan wound down. But, he said, "the world situation changed," with the rapid emergence of the Islamic State, and the demand for the drones shot up again.

Officials say that since August, Predator and Reaper drones have conducted 3,300 sorties and 875 missile and bomb strikes in Iraq against the Islamic State.

What had seemed to be a benefit of the job, the novel way that the crews could fly Predator and Reaper drones via satellite links while living safely in the United States with their families, has created new types of stresses as they constantly shift back and forth between war and family activities and become, in effect, perpetually deployed.

"Having our folks make that mental shift every day, driving into the gate and thinking, 'All right, I've got my war face on, and I'm going to the fight,' and then driving out of the gate and stopping at Walmart to pick up a carton of milk or going to the soccer game on the way home — and the fact that you can't talk about most of what you do at home — all those stressors together are what is putting pressure on the family, putting pressure on the airman," Colonel Cluff said.

While most of the pilots and camera operators feel comfortable killing insurgents who are threatening American troops, interviews with about 100 pilots and sensor operators for an internal study that has not yet been released, he added, found that the fear of occasionally causing civilian casualties was another major cause of stress, even more than seeing the gory aftermath of the missile strikes in general.

A Defense Department study in 2013, the first of its kind, found that drone pilots had experienced mental health problems like depression, anxiety and post-traumatic stress disorder at the same rate as pilots of manned aircraft who were deployed to Iraq or Afghanistan.

Trevor Tasin, a pilot who retired as a major in 2014 after flying Predator drones and training new pilots, called the work "brutal, 24 hours a day, 365 days a year."

The exodus from the drone program might be caused in part by the lure of the private sector, Mr. Tasin said, noting that military drone operators can earn four times their salary working for private defense contractors. In January, in an attempt to retain drone operators, the Air Force doubled incentive pay to $18,000 per year.

Another former pilot, Bruce Black, was part of a team that watched Abu Musab al-Zarqawi, the founder of Al Qaeda in Iraq, for 600 hours before he was killed by a bomb from a manned aircraft.

"After something like that, you come home and have to make all the little choices about the kids' clothes or if I parked in the right place," said Mr. Black, who retired as a lieutenant colonel in 2013. "And after making life and death decisions all day, it doesn't matter. It's hard to care."

Colonel Cluff said the idea behind the reduction in flights was "to come back a little bit off of 65 to allow some breathing room" to replenish the pool of instructors and recruits.

The Air Force also has tried to ease the stress by creating a human performance team, led by a psychologist and including doctors and chaplains who have been granted top-secret clearances so they can meet with pilots and camera operators anywhere in the facility if they are troubled.

Colonel Cluff invited a number of reporters to the Creech base on Tuesday to discuss some of these issues. It was the first time in several years that the Air Force had allowed reporters onto the base, which has been considered the heart of the drone operations since 2005.

The colonel said the stress on the operators belied a complaint by some critics that flying drones was like playing a video game or that pressing the missile fire button 7,000 miles from the battlefield made it psychologically easier for them to kill. He also said that the retention difficulties underscore that while the planes themselves are unmanned, they need hundreds of pilots, sensor operators, intelligence analysts and launch and recovery specialists in foreign countries to operate.

Some of the crews still fly their missions in air-conditioned trailers here, while other cockpit setups have been created in new mission center buildings. Anti-drone protesters are periodically arrested as they try to block pilots from entering the base, where signs using the drone wing's nickname say, "Home of the Hunters."

Source: http://www.nytimes.com/2015/06/17/us/as-stress-drives-off-drone-operators-air-force-must-cut-flights.html

Switzerland begins postal delivery by drone

By AGENCE FRANCE-PRESSE
The Guardian, 8 July 2015

BERN – Switzerland's postal service said on Tuesday it had begun testing parcel deliveries by unmanned drones, although widespread use of the flying postmen is not likely to kick in for another five years.

Postal service executives showed off the drones for the first time on Tuesday and said initial tests of the machines' post-delivery abilities would run until the end of July.

The snow-white drones consist of four branches with propellers on the end extending from a hollow ring the size of a toilet seat. A yellow box, bearing the postal service logo, is lodged in the middle.

"The drone has an extremely light construction and is capable of transporting loads of up to one kilo over more than 10 kilometers with a single battery charge," Swiss Post said in a statement.

The drone "flies autonomously, following clearly defined, secure flight paths, which are drawn up by cloud software developed by Matternet (the drone's US manufacturer)", Swiss Post added.

Swiss Post, which is cooperating on the project with Swiss WorldCargo — the air freight division of Swiss International Air Lines — stressed the drones would be thoroughly tested before being put to wide-scale use.

"Until the time of their realistic commercial use in around five years, there are various requirements which need to be clarified," the company said.

This includes exploring the regulatory framework that would apply when sending the unmanned aircraft out across the Alpine country, which is dotted with numerous remote and isolated villages where drone deliveries could be useful.

Swiss Post also said extensive tests would be carried out to explore the technical restrictions of the drones, including limited battery life.

For now, Swiss Post said it expected to mainly use the drones in emergency situations, which could "involve bringing supplies to an area that has been cut off from the outside world following a storm."

"Another realistic possibility is the urgent transport of consignments with the highest priority, such as laboratory tests," it added.

Switzerland is not the only place where package-delivering drones could soon appear. Amazon, the world's largest online retailer, announced in late 2013 a plan to airlift small parcels to customers by drone in select markets, less than 30 minutes after an order is received.

But the company warned last month that proposed US rules regulating the use of civilian drones could block it from launching the service, and called for them to be overhauled.

Source: https://www.theguardian.com/technology/2015/jul/08/swiss-post-begins-testing-postal-delivery-by-unmanned-drone

Mexico drug trafficking: drone carries 28 pounds of heroin across border to US

By SARAH BERGER
International Business Times, 13 August 2015

CALEXICO – Two men have pleaded guilty to smuggling drugs by a drone. This is the first cross-border drug seizure by US law enforcement involving a drone, federal authorities said Wednesday.

Twenty-eight pounds of heroin were smuggled by the drone across the US border near Calexico, near San Diego, in April. The two men who pleaded guilty were caught on Border Patrol cameras on April 28 retrieving the drugs near California Highway 98 in Imperial County, according to court records. US law enforcement officials called drug-carrying drones an emerging threat but questioned how profitable the new method of smuggling can be, because drones are limited in how much they can carry.

"With border security tight, drug traffickers have thought of every conceivable method to move their drugs over, under and through the border," said US Attorney Laura Duffy in a statement, the Los Angeles Times reported. "We have found their tunnels, their Cessnas, their Jet Skis, their pangas, and now we have found their drones."

This is not the first time drug smugglers have been caught using drones to transport their stash. In January, a drone carrying meth crashed in Mexico near the San Ysidro Port, KNSD-TV reported. It was likely being transported from neighborhood to neighborhood, according to Mexican law enforcement.

Small drones are able to fly for an hour and travel as far as five miles, and some can even be navigated by preset GPS coordinates. In recent years, drug cartels in Mexico have begun relying more on drones to transport their drugs. The Drug Enforcement Agency reported in July of last year that drug-carrying drones made an average of 150 trips between Mexico and the US, the Latin Times reported. Drones have been called the ideal "drug mules" because they allow drug cartels to transport shipments more quickly with less risk of being caught. Mexican media has reported that drug cartels have been commissioning engineers to design custom-made drones.

Source: http://www.ibtimes.com/mexico-drug-trafficking-drone-carries-28-pounds-heroin-across-border-us-2051941

New boss on construction sites is a drone

By WILL KNIGHT
MIT Technology Review, 26 August 2015

SACRAMENTO, CALIFORNIA – For some construction workers, any thoughts of slacking off could soon seem rather quaint. The drones will almost certainly notice.

The site of a lavish new downtown stadium for the Sacramento Kings in California is being monitored by drones and software that can automatically flag slow progress.

Once per day, several drones automatically patrol the Sacramento work site, collecting video footage. That footage is then converted into a three-dimensional picture of the site, which is fed into software that compares it to computerized architectural plans as well as a the construction work plan showing when each element should be finished. The software can show managers how the project is progressing, and can automatically highlight parts that may be falling behind schedule.

"We highlight at-risk locations on a site, where the probability of having an issue is really high," says Mani Golparvar-Fard, an assistant professor in the department of civil engineering at the University of Illinois, who developed the software with several colleagues. It can show, for example, that a particular structural element is behind schedule, perhaps because materials have not yet arrived. "We can understand why deviations are happening, and we can see where efficiency improvements are made," Golparvar-Fard says.

The project highlights the way new technologies allow manual work to be monitored and scrutinized, and it comes as productivity in other areas of work, including many white collar jobs, is being tracked more closely using desktop and smartphone software.

Software developed at the University of Illinois can show different stages of construction. Such additional scrutiny is sometimes controversial. The tracking of office workers raises worries over privacy, for instance, and fears that people may be encouraged to work excessive hours.

Golparvar-Fard concedes that this could be an issue, but he defends the idea. "It's not new to the construction industry that there would either be people standing and observing operations, or that there would be fixed cameras," he says. "Yes, making this autonomous has a different feeling for the workers. But you have to keep in mind that it's not really questioning the efficiency of the workers, it's questioning what resources these guys need to be more efficient."

Such concerns aren't slowing down development of the technology required for monitoring construction work. The falling cost of drone hardware and the availability of sophisticated control, navigation, and planning software have helped the aerial vehicles make a large impact on the agriculture industry already.

Monitoring activity across a large, complex construction site is particularly difficult because there are so many moving parts, and because the jobs being performed change frequently. A report published in 2009 by the National Research Council of the National Academies found that construction lags behind other industries such as manufacturing in terms of productivity, and blamed the situation on problems with planning, coördination, and communication.

At the Sacramento project, video is being captured by a drone-operating company called ImageInFlight. The software developed by the University of Illinois team can show how different subcontracting teams are working together.

Lincoln Wood, regional manager for virtual design and construction at Turner Construction, which is running the Sacramento project, says that while it is common to monitor progress closely, the near-real-time aerial images and software analysis being used there provides a more comprehensive picture of what's going on, and can highlight how a slowdown in one area may affect the entire project. "The nice thing about it is that it's showing all the tasks in an area, so people are seeing the global impact," he says.

The software developed by Golparvar-Fard and his colleagues including Timothy Bretl, an associate professor of robotics, and Derek Hoiem, an associate professor of computer science, is also being used at a high-rise construction project in Arizona, and by Taisei, a large construction company in Japan.

The University of Illinois team is currently testing a system that will allow drones to attach cameras to locations across a building site, so that activity can be monitored continually. In experiments, they are also using a crowd-sourcing platform to categorize workers' activities in video footage. A manager can then see how different tasks are being performed overall, and how much time each individual is spending on a job.

This level of surveillance could prove even more controversial. A spokesman for the Laborer's International Union of North America (LIUNA) said the organization was not aware of such technology, and declined to comment.

Source: https://www.technologyreview.com/s/540836/new-boss-on-construction-sites-is-a-drone/

Lord Norman Foster to build world's first droneport in Rwanda

By MADHUMITA MURGIA
The Telegraph, 21 September 2015

KIGALI – Lord Norman Foster, the British architect who has built iconic buildings like the Gherkin in London, is building the world's first "droneports" in Rwanda.

The goal is to transport urgent medical supplies and electronic parts to remote parts of the East African country via unmanned flying vehicles or drones.

"There will be about 2.2 billion people in Africa by 2050, or 1 in 4 inhabitants of the planet will be African. How can their infrastructure even think about keeping up with this expansion?" Lord Foster told the Telegraph.

The solution is clear: cargo drone routes have value wherever roads are limited. "Only a third of Africans live within 2 km of an all-season road," Lord Foster said. There are currently no continental motorways, almost no tunnels, and not enough bridges that can reach people living in far-flung areas of the continent.

The architecture firm Foster + Partners are working with Lausanne-based university École Polytechnique Fédérale de Lausanne (EPFL) and its associated initiative Afrotech on the project, which includes three separate droneports and is expected to be built within four years.

"[Afrotech director] Jonathan Ledgard approached me and said you've built the biggest airport in the world, now how do you feel about doing the smallest one?" Lord Foster said.

Each droneport will operate two parallel networks of drones, which are being designed by engineers at Afrotech/EPFL and Imperial College London.

The Redline drones will have a wingspan of 3m and can deliver medical and emergency supplies weighing up to 10 kilograms, such as 20 adult blood transfusions or vaccines and drugs.

The commercial Blueline drones can deliver heavier equipment, electronics and e-commerce items weighing up to 100 kilograms.

"For example, an entire village can be deprived of water because a vital component of a pump has broken down," Lord Foster says. "The drone can deliver in 1/12th of the time that it would take a Land Rover to get through."

The droneports will be designed as a row of vaulted brick structures and will host a health clinic, a digital fabrication shop to make spare drone parts, a post and courier room, and an e-commerce trading hub.

It is estimated that drones flying from the port will be able to cover 44 per cent of Rwanda.

Rwanda's challenging geographical landscape makes it an ideal test-bed for the pilot project, but it won't just stop there.

Lord Foster says he sees the potential for droneports to become as commonplace as petrol stations, and there are plans to expand the scheme across Rwanda and into neighbouring countries such as the Democratic Republic of Congo.

"This project can have massive impact through the century and save lives immediately," he said.

The Droneport project is an evolution of Foster + Partners' airport schemes in Mexico City, Jordan and Beijing, as well as a project to build a habitation on the Moon. The lunar project, in association with the European Space Agency, used a basic inflatable framework and sourced raw materials, like moondust, locally — very similar to the principles behind the droneports.

Lord Foster said, "What started as the world's smallest airport could become bigger than the biggest airport, in terms of sheer scale and impact."

> *Source: http://www.telegraph.co.uk/technology/news/11879956/Lord-Norman-Foster-to-build-worlds-first-droneport-in-Rwanda.html*

Turkey downs drone near Syrian border; Russia denies aircraft lost

By LOVEDAY MORRIS
The Washington Post, 16 October 2015

BEIRUT – Turkey said its warplanes shot down a drone Friday near its border with Syria, underscoring heightened hostility in the region's increasingly crowded airspace.

The aircraft was warned three times to turn back before being fired on by Turkish jets patrolling the border, the Turkish military said in a statement. It gave no further details.

Russia, which sent military planes into Syria last month, said none of its aircraft in Syria had been hit. But the United States suspected that Turkish fighters targeted a Russian drone, the Reuters news agency reported, citing a US official speaking on the condition of anonymity.

The incident occurred during a period of escalating tension along the Syrian border since Russia began an air campaign in support of President Bashar al-Assad 2 1/2 weeks ago. Turkey, a NATO member, has accused Russia of intentionally violating its airspace on two occasions since then. The incidents have drawn sharp rebukes from Turkey's NATO allies.

"Our aircraft performing patrol duty on the border fired and shot down the aircraft," the Turkish military statement said. It said Turkey acted "within the rules of engagement."

The drone went down near the village of Deliosman, in Kilis province, Turkish media reported. The province borders Syria's Aleppo province, where Syrian government forces, backed by Russian airstrikes, launched a new offensive Friday.

Turkish Foreign Minister Feridun Sinirlioglu said the drone crossed the border and flew nearly two miles into Turkey before it was shot down. Some of the wreckage was examined, and photos of the downed drone were circulated by the Turkish media.

The wreckage resembled that of a small Russian-made Orlan-10 reconnaissance drone, reported the Aviationist, a blog specializing in military aviation.

The United States, Russia and Syria operate drones in the border region.

In Moscow, Maj. Gen. Igor Konashenkov, a spokesman for the Russian Defense Ministry, said no Russian aircraft had been shot down. "All the airplanes of the Russian aviation group in the Syrian Arab Republic have returned after accomplishing their combat tasks," Konashenkov said.

In Lebanon, the pro-Syrian al-Mayadeen television channel echoed that denial, quoting a Syrian military official as saying that no Syrian or Russian warplane or drone had been shot down over Turkey.

Since 2013, Turkey has downed a Syrian military jet, a helicopter and a surveillance drone that crossed into Turkish airspace. Turkey announced that any aircraft entering its airspace from Syria would be considered a threat after Syrian ground fire brought down a Turkish F-4 Phantom II fighter jet in 2012.

The United States, which is leading separate airstrikes against the Islamic State, has raised concerns about possible confrontations between Russia and Turkey in the skies over Syria.

Defense Secretary Ashton B. Carter this month called on Russia to act in a "safe and professional manner" and respect Turkish sovereignty. NATO Secretary General Jens Stoltenberg said last week that the organization has "a duty to reinforce" its member state after the incursions, which he described as unacceptable.

NATO said Friday that Turkey has not asked it for help since the aircraft was downed.

"We understand that Turkish authorities are investigating the origin of the unidentified drone that was shot down. Turkey has not asked for any NATO assistance or consultations," said a NATO spokesman, who spoke on the condition of anonymity under customary ground rules.

In an attempt to ease friction with Turkey over previous warplane incursions, a high-level Russian military delegation held talks in Ankara on Thursday. Russia has blamed bad weather for what it says was a single incursion, and Turkey said it received apologies from the Russians.

"They apologized a few times, said it happened by accident and that they have taken measures so that it will not occur again," Sinirlioglu, the foreign minister, said following the meetings, Reuters reported.

Russian airstrikes have added new zeal to the beleaguered Syrian government's campaign. The Syrian army advanced to control four towns south of Aleppo on Friday, the state-run Syrian Arab News Agency reported. A US defense official and Syrian activists said this week that hundreds of Iranian troops were among those massing for an offensive near Aleppo. Pro-government forces have also launched attacks near the central city of Hama.

Michael Birnbaum in Brussels, Erin Cunningham in Cairo and Thomas Gibbons-Neff and Brian Murphy in Washington contributed to this report.

Source: https://www.washingtonpost.com/world/report-turkey-downs-drone-near-syrian-border/2015/10/16/c4e9a2c2-73f5-11e5-8248-98e0f5a2e830_story.html

Archaeologists plan on using cosmic ray muons to peer inside Egyptian pyramids

By EMIKO JOZUKA
motherboard.vice.com, 29 October 2015

CAIRO – Pyramids, the final resting places for ancient Egyptian pharaohs and queens, are pretty awesome to look at. But what's just as fascinating for archaeologists, engineers, and the public is getting to the bottom of a 4500-year-old mystery: how were they built?

In an attempt to answer that question, a team of researchers from Egypt, Canada, France, and Japan have joined forces under the aegis of the ScanPyramids Project, which is supported by the Egyptian Ministry of Antiquities.

The one-year project will kick off in November 2016, and aims to use everything from drone-mounted scanners, thermal imaging, and radiography to both produce a 3D model of the Khufu and Khafre's pyramids on the Giza plateau and the Bent and Red pyramids at the Dahshur necropolis, and to peer non-invasively inside them.

"We want to use these technologies to look through the stones and see if we can find rooms and secret passages behind the walls and inside the pyramids," Mehdi Tayoubi, founder of the French arts heritage non-profit HIP, told me.

The researchers will use muon tomography — powered by cosmic ray muons, subatomic particles that can penetrate almost anything — to see inside the pyramids. The tech was refined by researchers from Nagoya University in central Japan and the Japanese KEK particle physics research institute following the devastating Tōhoku earthquake and tsunami disaster on 11 March 2011. In Japan, it has already been used to look inside volcanoes and nuclear reactors.

Over the decades, several efforts have been made to uncover what's inside the pyramids as well as understand how they were built. In 1986, a non-invasive microgravimetry survey was conducted by a team sponsored by the EDF Foundation to measure the different density of the materials inside the pyramid, said Tayoubi. In 1999, French architect Jean-Pierre Houdin used 3D modeling to identify a construction anomaly, which gave rise to the "spiral structure" — a theory that suggested that pyramids were made from the inside.

Tayoubi explained that back then, researchers were able to find these spiraling structures inside the pyramids, but they were unable to say what exactly these anomalous structures were. "We still don't have an answer for what that structure is," noted Tayoubi.

In the following year, a team of Japanese researchers from Waseda University used ground-penetrating radars inside the corridors leading up to the queen's chamber in the Great Pyramid. The trick now, said Tayoubi, was to combine research forces and advance the usage of new and different technologies in heritage projects.

"When the Japanese team come to Egypt with their muon technology, they will meet with new challenges that will require them to find answers to new questions," said Tayoubi, noting that the project aimed both to advance innovations in tech, and unravel the mysteries of how Egypt's pyramids were built.

"I want the team to be in a humble position, as many people have tried before to work out the mystery of the pyramids, but we're really excited [about using these technologies]. If we find something, it's going to allow us to understand how the pyramids were built—this is one of the greatest mysteries," said Tayoubi.

Source: http://motherboard.vice.com/read/archaeologists-plan-on-using-cosmic-ray-muons-to-peer-inside-the-pyramids

Drone crashes onto piste, misses champion skier by inches

By MATIAS GREZ
cnn.com, 23 December 2015

Madonna di Campiglio, Italy – Defending World Cup champion Marcel Hirscher had a lucky escape Tuesday after he narrowly avoided being hit by a falling drone. The 26-year-old — who won silver in the slalom at the 2014 Sochi Winter Olympics — was on his second run in a World Cup slalom race at Madonna di Campiglio in Italy when a remote-controlled drone with a mounted camera slammed down on the piste inches behind him.

"This is horrible," Hirscher said after the event. "This can never happen again. This can be a serious injury."

He did, however, see a funnier side to the incident later that day, posting that there was "heavy air traffic in Italy" on his Instagram account.

Despite the near miss, the Austrian finished in second place, 1.25 seconds behind Norway's Henrik Kristoffersen.

The International Ski Federation (FIS) released a statement on its website apologizing for the "unfortunate accident." "While FIS and its partners aim to use new technology to enhance the fan experience, an accident such as the drone crash cannot happen again," the FIS statement read.

"Even if — unlike in Austria, Switzerland and other countries — drones are authorized to fly over a crowd during events in Italy, FIS and the host broadcaster will work together with all the involved parties to see what occurred during the crash and ensure that this will not happen again."

In the same statement, broadcast partner Infront Sports and Media also offered its apologies to Hirscher and the FIS.

"Detailed technical analyses will be conducted and more information will be given as soon as possible," they said.

Source: http://edition.cnn.com/2015/12/23/sport/marcel-hirscher-drone-crash/

SCDC

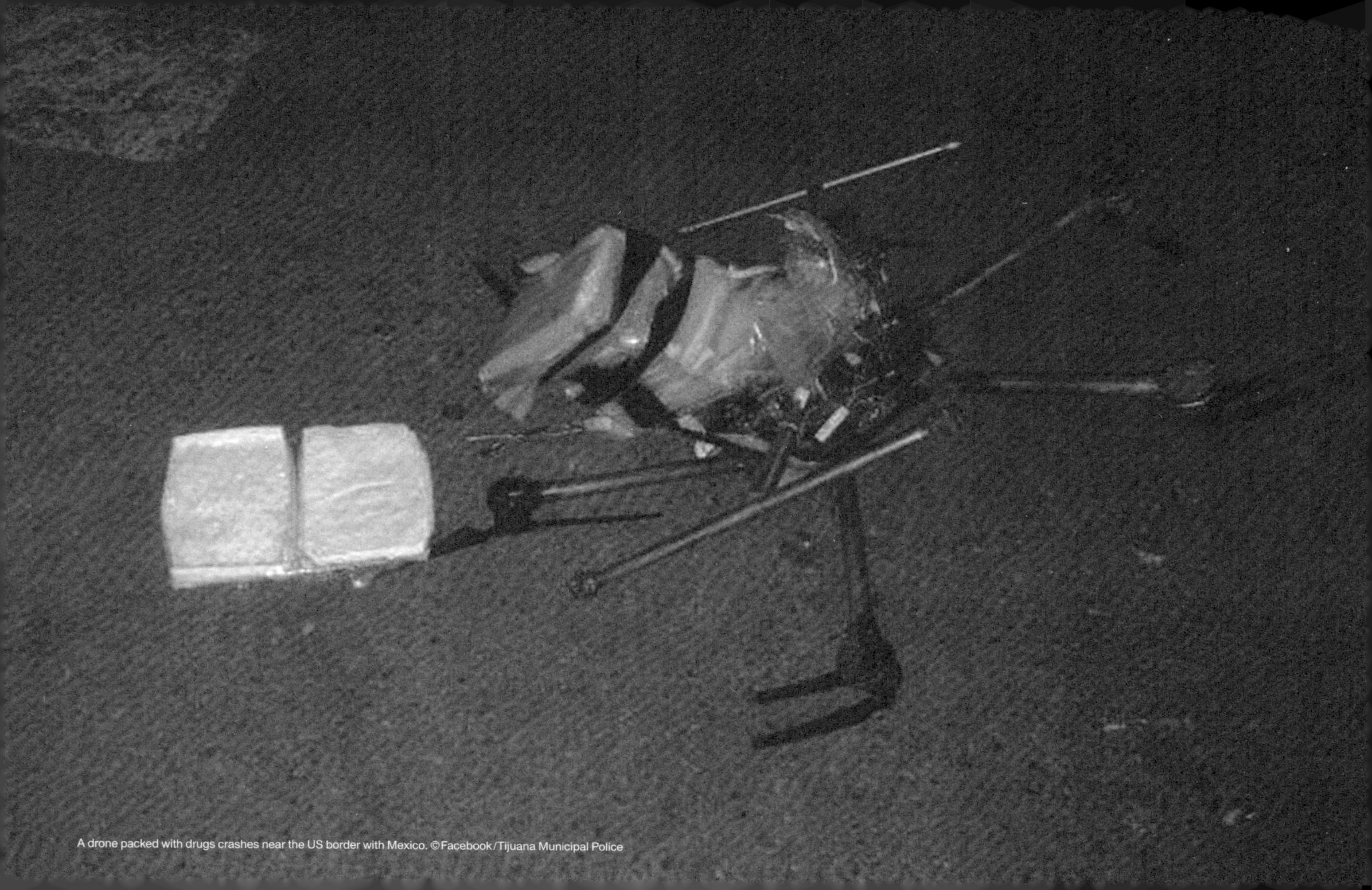

A drone packed with drugs crashes near the US border with Mexico. ©Facebook/Tijuana Municipal Police

A six-story Calvin Klein ad featuring Kendall Kenner got drone-tagged by artist KATSU. © ANIMALNewYork / Aymann Ismail

Three droneports designed by Lord Norman Foster will be completed in Rwanda by 2020. ©Foster + Partners

Marcel Hirscher evades a falling drone in a World Cup slalom race at Madonna di Campiglio, Italy. ©www.ruv.is

Drone swarms will change the face of modern warfare

By DAVID HAMBLING
wired.com, 7 January 2016

ARLINGTON COUNTY, VIRGINIA – The summer of 2016 will mark a significant moment in the future of air warfare: swarms of US Navy robots will leave the laboratory and be tested in the field. In the past, small numbers of extremely expensive manned aircraft were the norm, but, in the next few years, large numbers of cheap, expendable drones will be deployed in real-world situations. The US Navy's Low-Cost UAV Swarming Technology (LOCUST) programme is fusing unmanned aircraft into a swarm, and will demonstrate 30 drones flying together somewhere over the ocean.

The challenge won't be just getting them in the air at the same time — what makes a group into a swarm is its ability to coordinate. "A swarm operates as a unit," says Stephen Crampton, CEO of Swarm Systems, an autonomous-systems startup based in Hertfordshire. "It has a mission that it has to carry out, and it is self-reconfiguring so that if one drone gets taken out, the others autonomously change their behaviour to complete the mission."

Any aircraft, manned or unmanned, can be brought down by a single missile, but a swarm can take multiple hits and keep going.

Being able to absorb this type of damage makes the drone swarm hard to stop, especially when defences are geared towards shooting down single planes. Designing a system in which drones are aware of each others' movements is a big breakthrough. "It is the communication and/or awareness of each other that represents the major leap forward from remotely controlled and independent platforms to those that allow for collaborative behaviour," explains Lee Mastroianni, project manager of LOCUST. The programme has been working with the Georgia Tech Research Institute to develop a system in which individual drones will position themselves autonomously, flying in formation without being told explicitly where to go. "Precise formation control of large numbers of UAVs while conducting various manoeuvres is a major objective," says Mastroianni. Managing the swarm requires a new approach to control: instead of remotely piloting a single drone, the operator manages the swarm. He describes how the operator's interface will handle "aggregation" and "disaggregation", his terms for drones joining or leaving the swarm. A single drone might detach to get a closer look at a target, and return or carry out an attack.

The main hardware for LOCUST is the Coyote drone, a metre-long unmanned aircraft produced by defence manufacturer Raytheon. Coyote was designed as an expendable reconnaissance asset with folding wings, so that it can be fired from the tubes used for dropping sonar buoys on anti-submarine aircraft. Once clear of its launch tube, the Coyote's wings flick out and it can fly for up to 90 minutes on battery power, beaming back video from 30km away.

But the hardware is less important than the sensors and software enabling the drones to act in a swarm. "I am largely UAV agnostic," Mastroianni says. "The key is a modular UAVthat can easily acceptdifferent payloads depending on which missions are desired and can be produced cheaply enough that they are one-way."

Adaptability is important because different payloads are required for different types of mission: the drones may be equipped with video cameras or other sensors, jammers to interfere with enemy radar or they might carry explosive warheads for kamikaze-style attacks. This is why "low cost" is part of the name — the drones must be cheap enough to be expendable. "Reaching the point where an entire swarm of UAVs costs less than a single missile is the general objective," Mastroianni says. The goal seems realistic enough when you consider that the US Navy's Harpoon anti-ship missiles cost around $1.2 million (£770,000) each.

Offensively, the swarm can overwhelm adversaries: anti-aircraft batteries simply don't have enough missiles to stop them. And although the Coyote drones may be too small to sink a ship on their own, they could potentially knock out radar, missile launchers or other key systems, leaving them vulnerable to other attacks. In defensive mode, a swarm can form a protective cordon against fleets of fast boats like those used by Iran's Revolutionary Guard. The swarm might carry out high-risk reconnaissance missions, collecting imagery or other data from targets too well-defended for a Predator drone or a manned aircraft to approach.

Mastroianni says the biggest challenges for the swarm are not technical, but more based on perception: safety policies treat unmanned aircraft as if they are manned, meaning that they are highly regulated. And there is the added complication that the drones are largely autonomous rather than being individually remotely piloted, so they are not under human control at all times, like current drones, which makes policymakers cautious. "Establishing trust in autonomous UAV systems is not only the biggest challenge, but a major objective," Mastroianni says. Swarms at sea are a start, but the real impact will be when they engage in land warfare. Stephen Crampton, CEO of Swarm Systems, says the cluttered environment where drones have to avoid trees, buildings and power lines is far more difficult than open water. Autonomous sense-and-avoid for small drones is still in its early stages, but as processors get more powerful, it is becoming more reliable. Crampton says that other advances such as deep learning and neural networks also offer potential solutions and the technology is advancing rapidly.

Swarm Systems demonstrated a swarm of quadrotors at a Grand Challenge event for the Ministry of Defence in 2008. But the first step into the market is an individually-piloted drone called Nano — Crampton compares it to "flying binoculars". "When you've got a squad that is pinned down, they need flying binoculars, which they can send over the hill or around the corner," he says

British soldiers already have Black Hornet, a palm-sized helicopter with cameras, but it is useless in strong winds; Crampton claims the Nano will be able to fly 99 per cent of the time. The company's background makes the Nano a first step towards tactical swarms of many drones working together. "You can see soldiers having a peripheral reconnaissance swarm," Crampton says.

Battery life is a big issue for small drones. But a swarm can have a "hive", a base station where individual drones return for recharging while the rest continue their mission. To the operator, unaware of charging going on in the background, the swarm's endurance is unlimited. This approach is relatively easy for fixed bases; Crampton says a mobile hive for soldiers on patrol is more challenging. A walking robot such as Boston Dynamics' BigDog, however, offers a good starting point.

Next year will be a breakthrough year for drone swarms — once trust in swarms is established, development is likely to be rapid. Then Crampton's "peripheral reconnaissance swarm" will change the face of the battlefield — especially when the drones are armed.

Source: https://www.wired.co.uk/article/drone-swarms-change-warfare

Nearly 300,000 civilian drones registered in US in 30 days

By SAMUEL GIBBS
The Guardian, 26 January 2016

WASHINGTON – Nearly 300,000 drones have been registered in the US in the last 30 days, the US Federal Aviation Administration (FAA) has revealed.

Compulsory registration of civilian drones in the US was introduced on 21 December covering all manner of remote control flying systems from toys to aerial cameras. Any small unmanned aircraft weighing between 250g and 25kg must be registered before being flown outdoors, and pilots must be aged 13 or older.

FAA administrator Michael Huerta said: "The registration numbers we're seeing so far are very encouraging. We're working hard to build on this early momentum and ensure everyone understands the registration requirement."

Owners who registered before 21 January had their $5 fee waived, while the online registration system is only open to recreational users. Drones must be registered every three years, and clearly marked with their registration number. Anyone who operated a drone before 21 December must register before 19 February to continue legally flying it.

High profile incidents involving drones have highlighted the need for safety and accountability. In 2013 one was flown towards German chancellor Angela Merkel, and in May last year there were attempts to fly over the White House, while in December one narrowly missed downhill skiing champion Marcel Hirscher on the slope in Italy.

Supplying the FAA with the name, address and email address of the owner is part of the registration process, which is intended to help prevent abuse and nuisance drones. Currently when a drone is captured it is difficult to trace back to an owner. The FAA hopes that registration will help.

US transportation secretary Anthony Foxx said: "The National Airspace System is a great resource and all users of it, including [unmanned aerial system] users, are responsible for keeping it safe."

Around 1m drones were expected to be sold in the US by the end of 2015, which means only about one-third of those newly acquired drones have been registered.

The current legislation in the UK is different. It is illegal to fly a drone within 50m of a building or a person and 150m of a built-up area. In addition, the maximum flight height is 400m and the drone has to remain in line of sight and within 500m of the pilot. But civil drones do not currently need to be registered. Commercial drone pilots must complete a training course and apply for a permit- from the Civil Aviation Authority to fly the drone.

Source: https://www.theguardian.com/technology/2016/jan/26/300000-civilian-drones-registered-in-us-compulsory

Project Skybender: Google's secretive 5G internet drone tests revealed

By MARK HARRIS
The Guardian, 29 January 2016

TRUTH OR CONSEQUENCES, NEW MEXICO Google is testing solar-powered drones at Spaceport America in New Mexico to explore ways to deliver high-speed internet from the air, the Guardian has learned.

In a secretive project codenamed SkyBender, the technology giant built several prototype transceivers at the isolated spaceport last summer, and is testing them with multiple drones, according to documents obtained under public records laws.

In order to house the drones and support aircraft, Google is temporarily using 15,000 square feet of hangar space in the glamorous Gateway to Space terminal designed by Norman Foster for the much-delayed Virgin Galactic spaceflights.

The tech company has also installed its own dedicated flight control centre in the nearby Spaceflight Operations Center, separate from the terminal.

Based out of the site near the town called Truth or Consequences, Project Sky-Bender is using drones to experiment with millimeter-wave radio transmissions, one of the technologies that could under-pin next generation 5G wireless internet access. High frequency millimeter waves can theoretically transmit gigabits of data every second, up to 40 times more than today's 4G LTE systems. Google ulti-mately envisages thousands of high altitude "self-flying aircraft" delivering internet access around the world.

"The huge advantage of millimeter wave is access to new spectrum because the existing cellphone spectrum is over-crowded. It's packed and there's nowhere else to go," says Jacques Rudell, a pro-fessor of electrical engineering at the University of Washington in Seattle and specialist in this technology.

However, millimeter wave transmissions have a much shorter range than mobile phone signals. A broadcast at 28GHz, the frequency Google is testing at Spaceport America, would fade out in around a tenth the distance of a 4G phone signal. To get millimeter wave working from a high-flying drone, Google needs to experiment with focused transmissions from a so-called phased array. "This is very difficult, very complex and burns a lot of power," Rudell says.

The SkyBender system is being tested with an 'optionally piloted' aircraft called Centaur as well as solar-powered drones made by Google Titan, a division formed when Google acquired New Mexico startup Titan Aerospace in 2014. Titan built high-altitude solar-powered drones with wingspans of up to 50 meters.

Emails between Spaceport America and Google project managers reveal that the aircraft have exclusive use of the Space-port's runway during the tests and will even venture above the neighbouring White Sands Missile Range.

Google spent several months last summer building two communication installations on concrete pads at Spaceport America. Project SkyBender is part of the little-known Google Access team, which also includesProject Loon, a plan to deliver wireless internet using unpowered bal-loons floating through the stratosphere.

One of the millimetre wave transceivers was located near Spaceport America's Spaceport Operations Centre (SOC), and the other four miles away at the Vertical Launch Area (VLA), although Google's plans did not involve any rockets. Google also established a repeater tower and numerous other sites around the Space-port, presumably to test millimetre wave reception.

Both installations have cabinets full of computer servers and other electronics, while the pad at the SOC required a concrete base to support a dish antenna nearly eight feet across, according to a separate filing with the Federal Communications Commissions (FCC).

Work did not proceed smoothly, however. At one point in late August, a lorry showed up at 10.30pm, causing the Spaceport America team to complain to Google: "We have no loading dock and no means to remove a pallet … from the middle of the truck." The lorry was turned away without making its delivery.

Later, components were installed upside down or supplied by Google without the necessary shelves, nuts and bolts. Near the end of the build in October, Mike Bashore, information systems manager at Spaceport America, even emailed to his Google contact, "We want to run out to Home Depot for grounding straps." These are needed to protect sensitive electronics from static electricity. The nearest Home Depot hardware shop is over 100 miles from the Spaceport.

Google is not the first organisation to work with drones and millimetre wave technology. In 2014, Darpa, the research arm of the US military, announced a program called Mobile Hotspots to make a fleet of drones that could provide one gigabit per second communications for troops operating in remote areas.

Google has permission from the FCC to continue its tests in New Mexico until July. Spaceport America will be glad of the $300,000 SkyBender tests, as Virgin Galactic virtually mothballed its terminal following the 2014 crash of its prototype SpaceShipTwo vehicle in California. Christine Anderson, chief executive officer of Spaceport America, admits that the facility is now running out of money.

"We are transitioning to supporting all aspects of the spaceport from our operational budget, as the [state] bonds have been spent except for the amount reserved for the southern road," she wrote in a blog post earlier this month. "We are asking the legislature for $2.8m … We appreciate that our request is a lot of money, but we also feel that it is a relatively small amount to protect the state's $218.5m investment already made in the new and exciting commercial space industry."

Google is paying Virgin Galactic $1,000 a day for the use of a hangar in the Gateway to Space building, but had to split its SkyBender tests into two separate flight campaigns to ease Virgin Galactic concerns. An unnamed Virgin Galactic executive emailed Anderson before the tests to note: "We will be arranging numerous activities around these occupancy periods, which would be impacted if there was any [timing changes]." Google also had to promise not to take any photographs inside the building.

Anderson expects Virgin Galactic to unveil its new SpaceShipTwo at the Spaceport in February, and to begin flights there in 2018.

Google declined to comment.

Source: https://www.theguardian.com/technology/2016/jan/29/project-skybender-google-drone-tests-internet-spaceport-virgin-galactic

Eagles vs drones: Dutch police eye birds to prey on drones

By THE ASSOCIATED PRESS
680 News, 2 February 2016

The Hague, Netherlands – It may sound like a flight of fancy, but Dutch police are considering using birds of prey to swoop down and pluck rogue drones out of the sky.

Police are working with a Hague-based company that trains eagles and other birds to catch drones to investigate whether the birds can be used above large events or near airports, where the small flying machines are banned.

Dennis Janus of the national police said Tuesday that trainers exploit the birds' natural instincts to tackle the high-tech problem of drones flying in restricted areas.

Janus says the birds are trained "to think drones are their prey" and get a reward if they catch one.

Video released by police showed a small white drone with four propellers rising into the air and a bird of prey grabbing it from above with its talons in one fluid motion.

Dutch police will likely make a decision later this year whether to use the birds.

Janus said part of the evaluation includes research by a respected Dutch scientific organization into whether catching drones could harm the birds of prey.

Source: https://www.680news.com/2016/02/02/eagles-vs-drones-dutch-police-eye-birds-to-prey-on-drones/

Drones could follow trails to find lost hikers

By BEN COXWORTH
newatlas.com, 10 February 2016

Zürich – It's becoming increasingly likely that in the not-too-distant future, a robot may be what finds you if you're trapped in rubble at a disaster site. Now, it's also looking like a drone might come to your aid if you should get lost in the woods. That's because scientists have developed machine learning-based software that *already* allows quadcopters to follow forest paths better than humans.

The program was created by researchers at the University of Zurich, the Università della Svizzera italiana, and the University of Applied Sciences and Arts of Southern Switzerland.

In the course of its development, team members spent several hours hiking along trails in the Swiss Alps, taking over 20,000 photos with a helmet-mounted camera as they did so. The software analyzed these images, using a deep neural network to teach itself the distinct (and sometimes subtle) features that differentiate a trail from the surrounding environment.

The system was then used in a quad-copter, which was equipped with two video cameras for stereoscopic computer vision. When placed on a previously-unseen trail, the drone was able to autonomously orient itself and follow the trail with an 85 percent accuracy rate — humans, by contrast, scored 82 percent.

It is hoped that eventually, multiple drones could be combined with human search-and-rescue teams, allowing a greater area to be covered within the same amount of time. Additionally, the aircraft could be used to check hazardous trails, minimizing the risk to human searchers.

Before that can happen, however, the system needs to be developed further. In its current form, for instance, it's still not capable of identifying objects as being humans when it finds them.

Source: http://newatlas.com/path-following-rescue-drone/41766/

Drones are being trialled to speed-up HIV diagnosis in Africa

By MATT BURGESS
wired.com, 14 March 2016

Lilongwe, Malawi – Unicef is trialling the use drones to reduce waiting times for HIV test results in rural Africa.

The UN's children's charity has started testing a 10km flight route in Malawi, west Africa, with a custom made drone from Matternet.

The first test flight had the quadcopter travel from a community health centre to Kamuzu Central Hospital laboratory, while carrying a 'simulated sample' of a HIV test.

While still in testing, if the flights prove cost-effective the drones will be able to carry up to 250 tests, or 1kg of samples, at once.

Unicef said using drones in the diagnosis process had the potential to cut waiting times and could eventually be integrated into Malawi's health system. "The transport system is unreliable, it has got a lot of delays, bikes are expensive, they don't have money for fuel and results don't get picked up," Angela Travis, Unicef's chief of communication, advocacy and partnerships told WIRED. "The drone could solve a lot of those delays." "The samples are very light, they're on cards. They put the blood onto the cards and let them dry and then put them in a sealed bag."

Travis said the testing was only in its first stage but if the flights are successful health workers will be trained to complete the 20 minute flights and human samples will be incorporated into the tests. The tests have been approved by the country's health ministry and civil aviation authorities.

Around 17,000 children in Malawi suffer from HIV and in 2014 40,000 children in the country were born to HIV positive mothers. Mahimbo Mdoe, a Unicef representative in Malawi, said care for children who are born with HIV could begin earlier if the drones sped up the diagnosis process. "Quality care of these children depends on early diagnosis, which requires taking dried blood samples from the health centre to the central laboratory for testing," Mdoe said.

In February Unicef announced it was investing $9 million in open source technology projects that will be used to save children's lives. The UN body said the money, which is being provided by governments and investors, will be used to fund 60 early stage startups that will improve infrastructure, connectivity, power and transport in developing countries. "So far it has gone without a hitch," Travis said of the early drone tests in Malawi. She explained that the drones could also be used in a wider network to fly between health centres and collect multiple samples at once. "Quite a lot of the health centres are quite close to each other so there is also the thought that they could break the flights and hop between them, and gather the samples as they go," Travis said. "It has to be worked out but potentially it could be a bigger drone that can fly further and has a bigger battery."

Source: http://www.wired.co.uk/article/unicef-drones-hiv-malawi-diagnosis

British teen wins $250,000 in world's biggest drone race

By K.G ORPHANIDES
wired.co.uk, 14 March 2016

DUBAI – British teenager Luke Bannister has won the $250,000 (£173,900) first prize in World Drone Prix in Dubai, the world's biggest and most lucrative drone race, which has a total prize fund of $1 million (£696,700).

Flying for the Tornado X-Blades Banni UK team, Bannister, 15, beat 150 other teams, taking the prize in a tightly-steered nighttime race around the neon-lit outdoor circuit.

Decked out in racing overalls, pilots monitor their drones' position through VR headsets, which take a video feed from cameras mounted on the front of their drones.

Run by the World Organization of Racing Drones, the World Drone Prix rules call for participants, over the course of 12 laps, to "overcome a series of obstacles found on the aerial track incurring as few penalties as possible", "make a minimum of one Pit Stop per Race" and "finish first."

Teams must consist of at least four people, including a pilot, navigator, and technical crew, and must have a sponsor, but don't have to pay any fees to participate in the race. The roller coaster like circuit is 591m long, but that goes up to a total distance of 822m if you include both shortcut lanes and the extended-length 'joker lane'.

The World Drone Prix organisers have said that they're looking into locations for future racing events, saying that "many cities and countries have shown interest in hosting the next race. W.O.R.D. is currently evaluating the options based on region, local regulations and communities."

Meanwhile, Mohammed al-Gergawi, the United Arab Emirates' Minister for Cabinet Affairs told the Associated Press that Dubai is planning the World Future Sports Games, to take place in December 2017.

Events will include robotic swimming, running, wrestling and car racing, as well as drone flying. Al-Gergawi said that "we are trying to bring the future closer to us."

Source: http://www.wired.co.uk/article/british-teenager-luke-bannister-wins-worlds-biggest-drone-race

The numbers behind the worldwide trade in drones

By GEORGE ARNETT
The Guardian, 16 March 2015

STOCKHOLM, SWEDEN – There are more unmanned aerial vehicles (UAV) — or drones — being transferred between countries than ever before, according to data from the Stockholm International Peace Research Institute (Sipri).

Between 2010 and 2014, there were 439 drones exchanged compared to 322 in the five years previous to that.

Armed drones have become increasingly controversial in recent years following their use by the US in strikes against Pakistan-based militants. Critics say that even when specific targets are sought, far too many people are killed as collateral.

Although armed drones and airstrikes are most associated with the UAV debate — drones with the capacity to make armed strikes still make up a small fraction of overall trade — just 11 (2.5%) of the 439.

Armed drones were exported for the first time in 2007 when the UK received a delivery of two MQ-9 Reaper drones from the US, which were initially used in the Afghanistan conflict.

Last year, China became the second country to openly export armed drones by sending five to Nigeria for its fight with Boko-Haram. However, Sipri say there is also evidence that China sent some of the same CH-3 model to Pakistan in 2013.

This has prompted US to open up its own trade in armed drones to friendly and allied countries — previously it was just exporting to the UK.

The size of the trade: UK imports the most
In 2013, the US sent the huge (15-tonne) Global Hawk to Germany. It is the largest drone ever exported, as well as the most expensive ($130m) and the one with the biggest range of travel (14,000km). Orders have also been placed by Nato, Australia, Japan and South Korea.

Do not be mislead by this chunky unmanned vehicle though, the number of drones changing hands is not that sizeable. UAVs made up just 0.3% of total major arms trade between 2010 and 2014, according to Sipri.

There were 35 different countries importing UAVs in the 2010 to 2014 period, nine more than between 2005 and 2009.

On top of that list of importers was the UK, which took 55 drones from Israel on top of six further armed UAV from the US — which adds up to more than a third of global deliveries between 2010 and 2014.

So where do all these drones come from? Mostly Israel
The middle eastern country is the largest exporter of drones. Between 2010 and 2014, it delivered 165 units across the globe. The US came second with 132, followed by Italy's 37.

Since 1985, Israel has accounted for the majority (60.7%) of drone exports worldwide. Despite China's move into armed UAV exports, it only supplied 0.9% of total exports during this last couple of decades.

Source: https://www.theguardian.com/news/datablog/2015/mar/16/numbers-behind-worldwide-trade-in-drones-uk-israel

How airports and the drone industry are teaming up to protect planes

By MATT MCFARLAND
The Washington Post, 29 March 2016

HOUSTON – More than 50 US airports will test a new system to make themselves more aware of drones flying near their runways. Airport executives and the drone industry expect the Digital Notice and Awareness System (D-NAS) to improve safety amid concerns raised by 764 drone sightings near airplanes in 2015.

Drone hobbyists are required to notify airports of their plans when flying within five miles of an airport, but doing so has been a difficult process.

For one, hobbyists have to track down the appropriate phone numbers.

"You don't want to give out air traffic control's phone number on a public website for obvious reasons. Well, how are you going to manage that?" said Steve Runge, division manager of the Houston Airport System's safety and emergency management. "This solves many of the problems that we didn't know how to solve."

In some cases no one is even available to speak to at an airport.

"Especially on the eastern seaboard you're almost always within five miles of an airport no matter where you are. The amount of airports makes notifications very improbable if not impossible," said Justin Towles, staff vice president of regulatory and legislative affairs at the American Association of Airport Executives. "This system will allow that to become much easier and streamlined."

Drone operators using D-NAS will input the radius of their flight and how long they intend to fly using apps from drone manufacturers such as DJI, 3DRobotics and Yuneec, or the websites AirMap and some airports. This information will then be sent to operations staff and air traffic control at participating airports. By sending information this way, drone operators will be able to more easily notify these airports of their flight plans.

Proponents of the system believe that by simplifying the process, more drone operators will comply, which will make airports more aware and the skies safer for airline pilots and passengers.

"We think that people generally want to do the right thing, but they may not know what the right thing is," Runge said. His team jumped at the chance to participate in the pilot program after looking for ways to manage the increasing popularity of drones, like many airports.

Runge said an added benefit of the system will be the ability to see a live map of nearby drone flights, and to track usage over time. He plans to analyze the data and potentially add signs in areas near his airports where drone flights are common. Houston expects to have its system up and running in two weeks.

Down the road, Runge expects to work with the Federal Aviation Administration to deem certain areas around his three airports unsafe to fly drones in at all, most likely the paths where planes take off and approach runways. Runge envisions the system will eventually notify his airport and law enforcement officials automatically if a drone wanders into a forbidden area.

Ben Marcus, chief executive of AirMap, the start-up behind D-NAS, said that it had received near-unanimous interest from airports. There is no charge for drone hobbyists to use the system.

"Our main goal is to enable the future of drones," Marcus said. "We realize there are many elements, some of which are technology, which will have to be invented in order to facilitate that future."

Source: https://www.washingtonpost.com/news/innovations/wp/2016/03/29/how-airports-and-the-drone-industry-are-teaming-up-to-protect-planes/

Cloud-seeding drone makes first flight over Nevada

By ANDREW DALTON
engadget.com, 5 April 2016

HAWTHORNE, NEVADA – A group of researchers and aviators in Nevada have successfully completed the first flight of an unmanned cloud-seeding aircraft, the state's Desert Research Institute reports.

The flight took place last week at an FAA-approved drone test site in Hawthorne, Nevada, where the DRI has been exploring the possibility of using of unmanned aerial vehicles to address the state's persistent drought and other natural disasters through weather modification.

"We have reached another major milestone in our effort to reduce both the risks and the costs in the cloud seeding industry and help mitigate natural disasters caused by drought, hail and extreme fog," said Mike Richards, President and CEO of Drone America, which provided the aeronautics expertise to get the aircraft off the ground.

During an 18-minute flight, the "*Sandoval Silver State Seeder*," as they're calling the 11-foot-wingspan aircraft, successfully demonstrated its ability to deploy two silver-iodide flares which are commonly used to induce extra rainfall from clouds. While there's still some controversy over whether cloud seeding actually works, the *Seeder* only reached an altitude of 400 feet during the test run and therefore can't quite claim to be controlling the weather just yet. Regardless, the project's lead scientist called it "a tremendous accomplishment for the state of Nevada and everyone involved." There's even an inspiring promotional trailer.

The next steps for the research team will be to conduct actual cloud seeding tests around DRI's operations area in the Lake Tahoe basin.

While the thought of unmanned vehicles controlling the weather may be a little too close to Skynet for some, the practice of cloud seeding has been used in the past for innocuous purposes like clearing the air before sporting events. The project's chief engineer Amber Broch described it in a more optimistic way, saying the flight "also shows the tremendous potential to use unmanned systems as tools for environmental science and innovative natural resource applications."

Source: https://www.engadget.com/2016/05/04/cloud-seeding-drone-makes-first-flight-over-nevada/

Electric 18-rotor Volocopter makes first manned flight

By NICK LAVARS
newatlas.com, 8 April 2016

KARLSRUHE, GERMANY – The pursuit of personal flight took a promising, and kind of wacky, step forward today with the electric 18-rotor Volocopter lifting somebody into the air for the very first time. The maiden manned flight went off without a hitch, with the makers intent on using the emission-free aircraft as flying taxis a little further down the road.

Part multicopter, part helicopter and part alien spaceship, e-volo's Volocopter can take off and land vertically and also hover in the air when need be. Built from fiber composite material with a 450 kg (992 lb) take-off weight, nine individual batteries power the all-electric motors and 18 rotors to make for a top speed of 100 km/h (62 mp/h).

Despite its complex shape and daunting appearance, the Volocopter is designed to bring personal flight to the masses by being easy to fly — so easy that anyone can apparently have a go. But for now, that privilege has been extended only to e-volo's managing director Alexander Zosel, who took grabbed hold of its single joystick and took the Volocopter to the skies over an airfield in southern Germany on March 30.

"It is definitely a sublime feeling to lift off, fly the first few meters, and then actually take my hand off the joystick and think that, yeah, it's really as if I'm standing on the ground, and then I look down and there are 20-25 meters beneath me," Zosel said. "So it's definitely unbelievable what we've achieved here. It's seriously unbelievable!"

Zosel's test flight saw the Volocopter hit speeds of 25 km/h (15 mph) at only low altitudes. This marks the first part of e-volo's test program. The second phase will test the aircraft at speeds of 50 km/h and a medium altitude, with the final phase to raise that to higher altitudes and its top speed of 100 km/h.

The makers have carried out more than 100 test flights in the buildup to this achievement, though these were unmanned and controlled remotely. The makers say that through this testing they have demonstrated the aircraft's ability to fly almost autonomously.

If things continue to run as planned for e-volo, it plans to start building large quantities of the Volocopter and enter the market within the next two years, with a view to eventually offering air taxi services and another form of public transport. They are likely to have some competition, however, with a number of other personal flight machines making promising strides of late. These have included the Malloy Aeronautics Hoverbike, Ehang's 184 AAV personal drone and the Snowstorm electric multicopter, built by students in Singapore.

Source: http://newatlas.com/volocopter-manned-flight/42704/

Drone hits plane at Heathrow airport, says pilot

By KEVIN RAWLINSON
The Guardian, 17 April 2016

London – Police are investigating a pilot's claim that his plane was struck by a drone as it approached Heathrow airport.

The Metropolitan police said they were contacted on Sunday afternoon by the pilot, who landed the plane safely at Terminal 5. No one has been arrested, officers said.

The flight, BA727, was coming in to London from Geneva, carrying 132 passengers and five crew.

British Airways said the Airbus A320 had been examined by engineers and cleared to take off for its next flight after the incident.

Steve Landells, flight safety specialist at the British Airline Pilots Association (Balpa), said: "Frankly it was only a matter of time before we had a drone strike given the huge numbers being flown around by amateurs who don't understand the risks and the rules."

"It appears that no serious damage was done on this occasion, but what is clear is that while most drones are flown safely, sensibly and within the limits of the law, much more education of drone users and enforcement of the rules is needed to ensure our skies remain safe from this threat."

Earlier this month, the airline pilots' union called for an investigation into the likely effects of a drone strike on an aircraft after a report by the UK Airprox Board found that there were 23 near-misses between drones and aircraft in the six months between April and October last year.

They included one on 22 September, when a Boeing 777 that had just taken off reported that a drone narrowly passed the right hand side of the airliner. Investigators concluded that the drone was at the same height and within 25 meters of the jet. A report was made to police but the drone operator was not traced.

Days later, a drone was flown within a few meters of an Airbus A319 landing at Heathrow. The pilot told the UK Airprox Board the drone may have been just 20 feet above and 25 yards to the left when it passed by the aircraft.

The jet was flying at an altitude of 500 feet and was on the final approach when the drone was spotted.

After that incident, Landells had discussed the possibility of engine failure in the event of a drone striking a plane. "You end up with very high-velocity bits of metal going anywhere they like. That could be through fuel tanks, through hydraulic lines and even into the cabin," he said.

"Losing the engine is not going to cause an aircraft to crash because they are designed to fly with one engine down. But an uncontained engine failure is going to be different every time. That could be very serious indeed."

Regarding the call for an investigation into the likely effects of a drone strike, Landells said: "The first thing we want to do is get a drone or at least the critical parts of a drone flying at a windscreen of an aircraft. The indications so far with computer modeling are that you'll end up with penetration of a windscreen."

"One possibility is that the battery smashes the windscreen and the inside layer of the windscreen shatters and you end up with a lot of glass in the cockpit, probably moving at quite high speed."

"As a pilot, I don't want to be sitting there when that's going on."

Source: https://www.theguardian.com/uk-news/2016/apr/17/drone-plane-heathrow-airport-british-airways

Drone delivers abortion pills to Northern Irish women

By PRESS ASSOCIATION
The Guardian/PA, 21 June 2016

BELFAST – Pro-choice activists have delivered abortion pills to women in Northern Ireland using a drone. The stunt aimed to highlight the strict laws around terminations on both sides of the Irish border.

Courtney Robinson, 18, from Belfast, who took the tablets, said: "We are here to say we are going to defy the law in helping women obtain these pills and we are going to work to make the law unworkable and stand in solidarity with all women who want to have an abortion and have the right to do so in Northern Ireland."

The drone flight started at Omeath in County Louth in the Republic and landed a short distance away, near Narrow Water Castle in County Down shortly after 10am on Tuesday.

Robinson, a member of Labour Alternative, said: "The reason we are doing this is to highlight that these pills are available to women who are not able to travel outside of Northern Ireland for an abortion. I have no concerns. I know the pills are safe."

"As long as politicians in Stormont and the Dáil [the Irish parliament] continue to ignore human rights we will continue our campaign." Uniformed and plain-clothes police officers at the landing site spoke to organisers and filmed the events, however no action was taken to confiscate the medication, which had been prescribed by a doctor. Traffic was also stopped on the main road while the drone was in the air.

The drugs, Mifepristone and Misoprostol, can be taken up to nine weeks into a pregnancy and have been approved for use by the World Health Organisation since 2005.

Lucy Simpson, from Belfast, who also took tablets, said legal reform was urgently needed. She said: "The law is archaic. We are governed in Northern Ireland by an act which is dated 1861, which is in the dark ages, it's like when dinosaurs were on earth. We think it should be changed radically and we can't really wait any longer."

"Thousands of women suffer every year in Northern Ireland and the Republic having to travel abroad for abortions and go through very traumatic times. We feel now is the time to change legislation." The event was organised by a collaboration of pro-choice groups, Alliance for Choice; Rosa; Labour Alternative and Women On Waves, which staged a similar flight from Germany to Poland.

The groups said no laws had been broken, adding: "The 'abortion drone' will mark the different reality for Irish women to access safe abortion services compared with women in other European countries where abortion is legal."

In Northern Ireland, the maximum penalty for the crime of administering a drug to induce miscarriage under the relevant law — the Offences Against the Person Act 1861 — is life imprisonment.

In April a 21-year-old woman was handed a suspended sentence by a judge in Belfast after she bought drugs on the internet to induce a miscarriage because she could not afford to fly to England.

In the Irish Republic, the offence of procuring an abortion carries a potential 14-year jail term.

Later, campaigners plan to picket Belfast high court during an appeal against a ruling which found that the law in Northern Ireland was incompatible with human rights legislation.

Anti-abortion organisations had vowed to stop the drone, however there was no protest at Narrow Water.

Source: https://www.theguardian.com/uk-news/2016/jun/21/drone-delivers-abortion-pills-to-northern-irish-women

Iraqi militias are using consumer drones to fight Isis

By AUSTIN BODETTI
wired.com, 28 June 2016

BAGHDAD – The owner of a small toy store in al-Mansur, a Baghdad neighborhood five kilometers from the Green Zone, Abu Abdullah sells a variety of remote-control vehicles. Some of his recent customers have been Iraq's notorious Shia militiamen, eager to purchase the handful of hobbyist drones available. "They've come here to buy drones before," Abu Abdullah tells WIRED. "I don't know why. Those drones aren't particularly effective or useful. They're toys."

When Isis overran Mosul and Tikrit two years ago, Iraqis who had lost faith in their military rallied behind the Iranian-backed Shia militias. Lacking the advanced technology available to the American-led coalition and the Iraqi government, many militias have resorted to consumer drones to help with intelligence gathering.

While toys to some, the drones are playing a crucial role in the conflict according to Sadiq al-Husseini, a commander in the Badr Organisation — the largest and oldest of Iraq's Shia militias. "The drones have been extremely useful for preventing casualties among our forces," says al-Husseini while observing the front lines around Fallujah, a former Isis stronghold. "They have helped us lock onto targets with our mortars and cut Isis's supply lines."

Badr and other Shia militias provided fire support to the Iraqi Security Forces, namely the Federal Police, the Iraqi Army and the Iraqi Counterterrorism Service, as they retook and secured Fallujah. Several militiamen explained that the drones assisted their artillery batteries in shelling Isis bases deep inside the city without hitting civilians, whom Isis had used as human shields. Considering the militias' reputations for war crimes, their bombardment of Fallujah remains controversial.

Even the army and police, both of whom receive equipment and other support from Western countries, have relied on hobbyist drones for similar purposes. The website of Kataib Sayyid al-Shuhada, another Shia militia, shows an Iraqi soldier deploying a DJI Phantom drone.

On a trip to the front lines in Fallujah organised by Badr, WIRED observes a contractor for the Ministry of Interior deploying a small drone from the rooftop of a three-story building to reconnoiter and surveil Isis positions deep in the city.

According to the contractor, Hussam al-Mayali, the drones' batteries only last forty minutes before the operators have to recharge them, so there are always spares on hand for longer missions — al-Mayali's deployment lasted ten minutes. The Ministry of Interior started with three such drones in Fallujah, but Isis destroyed one and the American-led coalition — likely by accident — hit another. "These drones allow us to watch Isis positions and call in artillery strikes on precise locations," says al-Mayali.

"Our artillery observers have brought our accuracy from 70 per cent to 95 per cent," explains Karam Hatam, a policeman manning a truck-mounted rocket launcher in the rear. Two rockets from the police's artillery batteries had hit Fallujah during al-Mayali's deployment of the drone.

"We use the drones to provide oversight of the battlefield as our soldiers advanced against Isis," says brigadier Haydar al-Kaabi, a commander from the Ministry of Interior. "Our logisticians buy them at the markets in Baghdad, where the militiamen purchase them like all other Iraqis."

Al-Husseini refused to describe how and where the Shia militias acquired the drones, only saying that the prime minister's office obtained them on the militias' behalf. Since 2014, all the Shia militias have reported to the prime minister's office. Badr prefers the American, Chinese, and German models for military missions, adds Al-Husseini.

In theory, the police should benefit from the reconnaissance and surveillance conducted by the drones of the American-led coalition as the army and the counterterrorism service have. But coordination between the branches of the Iraqi security forces remains poor, and the Ministry of Interior has ingratiated itself with the Shia militias at the expense of its relationships with the army, the counterterrorism service and the Americans. Further adding to the problem, the role of Minister of Interior has for the past two years belonged to Mohammed Al-Ghabban from Badr's eponymous political party.

In mid-June, the Institute for the Study of War reported Badr and the police had entered Fallujah together (despite the prime minister ordering the Shia militias to stay on the outskirts of the city) and that, concerning human rights defenders, the two had abused civilians in tandem.

Unlike soldiers from the army and counterterrorism service interviewed by WIRED, the police expressed little interest in working with the American-led coalition and its high-tech drones. "We don't trust the drones of the international coalition," says brigadier al-Kaabi. "America would provide Isis with advanced drones before it would help us." For now, the Ministry of Interior and most of its allies in the Shia militias must get their drones at toy stores such as Abu Abdullah's.

Iran has equipped some of its Iraqi proxies, including Badr, with unarmed but military-quality drones. In December 2012, the Iranian military claimed it had captured an American copy of the Boeing Insitu ScanEagle — a long-endurance drone for reconnaissance and surveillance based on a commercial drone intended for spotting fish. A few Iranian variants of the ScanEagle, such as the Yasir, have appeared in the hands of Iraqi Shia militias on battlefields in Fallujah and elsewhere.

Commanders from Kataib Hezbollah, a Shia militia that the US State Department has labelled a terrorist organisation for killing Americans during the Iraq War, confirmed that they are using Iranian drones in front lines near Makhul and Saqlawiya.

"Over the past few years, Iran has exported drones to a number of non-state actors, including Hezbollah and Hamas," says Dan Gettinger from the Centre for the Study of the Drone at Bard College in New York. "I can't say why exactly Iran would distribute this particular technology, but I can speculate that it seeks to test the drones in a combat environment." Hamas and Hezbollah have also prepared drones to subvert the overwhelming air supremacy of the Israeli Air Force's well-trained pilots.

In addition to conducting surveillance through Hamas and Hezbollah, Iran has deployed its own drones to Iraq and Syria, sometimes launching airstrikes. "When the targets are in range of Iran's borders, the army's ground forces haven't been shy about operating their 'border security' UAVs over Iraqi airspace," Galen Wright, an expert on Iranian drones tells WIRED. Of the Iraqi Shia militias' Iranian models, Wight says it is "unclear to what degree these are actually operated by Iraqis and to what degree they are accompanied by [Iranian] technical advisors."

The use of airborne terrorist threats, though recent, has precedent. During the later stages of Sri Lankan Civil War, the Tamil Tigers formed a nascent air force: the Air Tigers. The Tamil Tigers, already notorious for suicide bombings, sent their pilots on a Kamikaze-style raid. A Palestinian journalist explains that Hamas has been developing related 'suicide drones.'

Drones have spread to the hundreds of anti-government paramilitaries in Syria, where journalists have noted anecdotal reports of Isis developing its own models to carry explosives. The terrorist organisation has already used drones for reconnaissance and surveillance in Iraq. "Some factions of the Syrian opposition have drones, but they're sparsely used," says captain Rashid al-Hourani, leader of the Homs Liberation Movement. "The drones assist the revolutionaries in the tactical development of their plans, causing the bases of the [Syrian] regime to fall to them. They're very useful" Syria's rebels, adds Al-Hourani, procure the drones off the internet.

In Iraq, the Shia militias have allied themselves with the country's American-backed government, meaning that, at least for now, they have little reason to restart the war against Western soldiers stationed there. Even so, the Shia militias' access to this technology still requires scrutiny. "The drones locate areas of Fallujah where there are large concentrations of civilians to guide the militias' rockets toward them," Muhammad al-Issawi, a civilian in Fallujah, tells WIRED via Facebook Messenger in reference to Kataib Hezbollah. In other words, the militias were using the drones to commit war crimes in Fallujah.

With drones from Baghdad's commercial markets and Tehran's military factories, Iraq's Shia militias may once again catch the attention of Western governments. The US Department of Homeland Security has already voiced its alarm at the potential spread of hobbyist drones to terrorists.

"Generally speaking, there's space to be worried about proliferating drone technology because of the capabilities it bestows on the operator," observes Wright. "But the degree to which the technology is available commercially in one form or another means that it's more or less inevitable and that there shouldn't be too much hand wringing about it."

"It's always worrying when terrorist-labelled groups obtain a new technology," adds Gettinger. "If used effectively, drones could enable non-state actors to adopt more sophisticated tactics than before."

Source: http://www.wired.co.uk/article/iraq-isis-war-consumer-drones

Facebook's solar-powered internet plane takes flight

By SEAN FARRELL
The Guardian, 21 July 2016

YUMA, ARIZONA – Facebook has announced the first successful test flight of a high-altitude solar plane to bring internet access to remote parts of the world.

The Aquila drone has the wingspan of an airliner but weighs less than a car. When cruising it consumes just 5,000 watts — the same as three hairdryers or a powerful microwave.

The first flight took place on 28 June in Arizona. Facebook said the test went better than expected and that Aquila's 96-minute flight was three times longer than planned.

Aquila was developed in Bridgwater, Somerset, and the drone, designed to fly non-stop for three months, will use lasers to beam down internet access to remote areas without online capacity.

Facebook has a team of engineers at Bridgwater from fields of expertise including aerospace, avionics and software and who had previously worked at organisations such as Nasa, Boeing and the Royal Air Force.

Mark Zuckerberg, Facebook's chief executive, revealed in March 2015 that the company had been testing drones in the skies over the UK.

Facebook intends Aquila to be part of a fleet of planes that will provide the internet to 4 billion people in sub-Saharan Africa and other remote regions that do not have access currently.

Jay Parikh, Facebook's head of engineering and infrastructure, said in a blog: "We're encouraged by this first successful flight, but we have a lot of work ahead of us ... In our next tests, we will fly Aquila faster, higher and longer, eventually taking it above 60,000 feet."

> *Source: https://www.theguardian.com/business/2016/jul/21/facebook-solar-powered-internet-plane-test-flight-aquila*

Swedish church to use drones to drop Bibles on Isis territory in Iraq

By GABRIEL SAMUELS
The Independent, 9 August 2016

UPPSALA – An evangelical church in Sweden has announced it wants to airdrop thousands of copies of the Bible onto areas of Iraq controlled by Isis militants, in an effort to "pass the hope and love of the Christian gospel" to local people.

The Livets Ord church in Uppsala, north Sweden, is planning to fly drones at high-altitude over Isis-occupied regions, which will then drop tiny, electronic copies of the gospels into occupied towns.

The church, one of the leading charismatic institutions in Sweden, confirmed the Bibles are "the size of pill boxes" and "require no electricity but work on their own".

Representatives from the organisation dismissed accusations the Bible drop was intended to be provocative in any way, saying they were merely motivated by an "unfailing and never-ending love" for all people.

In a statement, the church said: "Our ambition is to pass on the hope and love of the Christian gospel to a population living in closed areas where they are being denied human rights."

"We start our project in a few weeks and hope to drop thousands of Bibles."

"We believe in an unfailing and never-ending love, given to us by God. And together we try to spread that love to as many people as possible."

The church has not yet outlined how precisely it will carry out the ambitious delivery, but remain convinced it will be possible despite having to fly through dangerous airspace. The operation is likely to place "in a few weeks", the church said.

Livets Ord, translated as 'The Word of Life', was founded in the 1980s and has frequently been criticised for being organised like a "cult", according to *Christian Today*.

> *Source: http://www.independent.co.uk/news/world/middle-east/swedish-church-to-use-drones-to-drop-bibles-on-isis-territory-in-iraq-a7181181.html#gallery*

Agriculture drones are finally cleared for takeoff

By PRACHI PATEL
IEEE Spectrum, 19 October 2016

WASHINGTON – Tech-savvy farmers have been some of the earliest commercial adopters of drone technology, purchasing 45,000 drones last year alone. But if they were using the drones to check on the condition of their fields, spraying their crops, or keeping tabs on livestock, most of them were technically breaking the law. New US federal rules that went into effect this summer, however, should make it easier for farmers to get a drone's-eye view of their fields.

The new rules allow commercial drone operators to get certified via a written test, so long as they fly drones that meet certain weight and altitude guidelines. Before this, operators had to pay for a pilot's license and get a special exemption to use a drone, a slow and cumbersome process.

The rules are a win for farmers, drone companies, and everyday consumers. Worldwide, agriculture is the largest commercial market for unmanned aerial vehicles, or UAVs. And it is expected to bag 80 percent of all commercial drone use in the United States. About half of the 5,500 exemptions that the US Federal Aviation Administration (FAA) has approved up to now were for agricultural purposes, says Brian Wynne, president and CEO of the Association for Unmanned Vehicle Systems International. With the new rules, the FAA expects more than 600,000 commercial UAVs to be flying during the next year. "These numbers foreshadow the exciting potential that UASs [unmanned aircraft systems] have for the agriculture industry now that the small UAS rule is in place," Wynne says.

By using drones to scout for weeds and pests, spot diseased plants or dry areas, and spray the right amount of fertilizer and pesticide, farmers can increase yield with less resources and environmental harm. Returns are especially steep for high-value crops like wine grapes. Drones are cheaper than hiring a small plane and, in contrast to satellites, work in cloudy conditions and give higher-resolution images. Prices for ready-to-fly agricultural UAV systems — hardware, sensors, and software included — range from US $1,500 to over $25,000.

"This is going to have a big impact on farmers," says Kyle Miller, an applications engineer at agricultural drone maker

AgEagle Aerial Systems, in Neodesha, Kan. Miller, who has been flying robotic aircraft over his family's soybean farms as a hobbyist for years, says interest in the technology among farmers he knows is soaring. "Now that the new rules are out, farmers know drones are coming to their area," he says.

Until now, many drone technology providers sat on the sidelines, not wanting to go ahead with drone testing or use given the lack of a clear regulatory path, says Nathan Stein, agricultural solutions manager at Swiss drone maker senseFly. "Whether it was a seed company or chemical company, their legal departments wouldn't let them move forward," he says. "Anything with the word drone in it was ominous."

Several US businesses have focused on operations in other countries. Drones have been used legally for agriculture for more than 20 years in countries such as Australia, Japan, and South Korea, according to Wynne. "Now companies can bring business and technology back into the United States," Stein says.

Plus, American farmers might reap more benefits from drones than farmers in other countries. "The United States has incredibly diverse agriculture, from peaches to corn to grapes to cotton," Stein says. What's more, he says, US farmers have "an adventurous spirit, so I expect a large market. People might come up with more creative ways to use drones on the farm."

With pressure mounting on the FAA from stakeholders, UAVs are poised to have an even bigger impact on agriculture. Current rules limit commercial drones to less than 55 pounds (25 kilograms), a maximum speed of 100 miles per hour (160 kilometers per hour), and an altitude of 400 feet (122 meters). They also must not fly above people and remain within the operator's visual line of sight.

That last rule is a killer for agriculture and other economic uses. Farms in the United States — think midwestern corn and soy expanses — are often much bigger than line-of-sight operation will allow for. There are around 2.1 million farms in the country. Of those, the 1.85 million or so smaller family farms (averaging 93 hectares) that make up about half the total US farm acreage can benefit from drones right now. The same isn't true for larger farms, which account for 36 percent of total US farmland.

The FAA is moving in the right direction, but it needs to do more, says Darryl Jenkins, chairman of the American Aviation Institute, an independent think tank for commercial aviation based in Washington, D.C. "Things move at a glacial speed at the FAA," Jenkins says. "To really get economic benefits from drones, we need beyond-line-of-sight operations, flights over people, and access to higher altitudes."

Perhaps then, like tractors and combines, drones might become standard farm equipment.

Source: https://spectrum.ieee.org/robotics/drones/agriculture-drones-are-finally-cleared-for-takeoff

Intel invented a way for a single operator to fly hundreds of drones at once

By APRIL GLASER
Recode, 4 November 2016

SANTA CLARA, CALIFORNIA – Intel unveiled a tiny new drone today, the Shooting Star, designed specifically for light shows in the night sky.

Weighing in at less than a pound, the small quadcopter is designed to fly in unison with hundreds of other drones, all controlled by a single operator.

An individual pilot flew 500 Shooting Star drones simultaneously in early October, breaking the Guinness Book of World Records for most drones operated at once by a single pilot. The flight took place in a small town outside of Munich, Germany, where Intel received a waiver from local regulators to perform the record-breaking flight.

Each drone is packed with full-color-range LED lights. The drones fly with the help of Intel's automation software, which allows artists to design aerial animations with hundreds of drones. It's the type of coordination that would typically take weeks or months, but Intel says its system can design a complex light show in days.

Intel is better known in the drone world for its collision avoidance camera technology, Real Sense, which is used in its industrial-grade Falcon 8 drone, as well as Yuneec's Typhoon H drone. The system allows drones to weave between obstacles without being piloted, but the new Shooting Star drones aren't equipped with Real Sense.

Instead, the drones are completely controlled by the master computer and operator on the ground. The drones don't communicate with each other either, so they're not a swarm, but are rather radio controlled.

Although the Shooting Star is the company's second fully-branded Intel drone — the first was the Falcon 8+ in October — neither is available for sale yet, and the company isn't sharing how it plans to bring the technology to market.

But Intel is better known for what it packs inside computers, like chips and microprocessors, and is less known for its hardware offerings. So until the company provides clarity about how others can buy and use its drones, the technology could just be a demonstration of what might one day end up in other machines, rather than Intel's own marketable product.

Shooting Stars can travel up to 22 miles per hour and can fly in light rain.

Source: https://www.recode.net/2016/11/4/13517550/intel-single-operator-fly-hundreds-drones-shooting-star

Man who used drone to deliver barbecue sausage to friend in hot tub could be fined

By JONATHAN PEARLMAN
The Telegraph, 9 November 2016

SYDNEY – Authorities in Australia are investigating a man who used a drone to pick up a sausage from a barbecue and deliver it to a friend in a hot tub.

The man faces a £5,600 [$AUS9,000] fine after he posted a video on YouTube which showed him sitting at home and piloting a drone which travelled over a busy motorway before descending to collect the cooked sausage from a barbecue — or "sausage sizzle" — at a hardware store in Melbourne. The video has since been removed.

Australia's Civil Aviation Safety Authority said it was investigating the incident over concerns the flight breached rules which ban drones from entering populous areas or flying within 33 yards [30 metres] of people.

"You can clearly see people walking to and from their cars, you can clearly see people around the sausage sizzle," a spokesman told Fairfax Media. "This is a classic example of a place where you should never fly a drone."

The video begins with a man writing a note which said: "Please buy a snag [sausage] and put in bag. Here's $10."

It then shows the drone flying to meet a man at the barbecue — apparently an accomplice — who put the sausage into a container attached to the machine.

The sausage was then delivered to a man in a hot tub in a back garden.

The man who posted the video online told EFTM, an Australian men's lifestyle website, he wanted to buy a new drone and had sold the footage to a viral video company.

He said the footage had been edited and the flight posed no threat to the public.

"We shot it in parts, never going over homes or people," the man said.

Source: https://www.telegraph.co.uk/news/2016/11/09/man-who-used-drone-to-lift-sausage-from-barbecue-and-deliver-to/

Boeing 'Base Station' concept would autonomously refuel military drones

By DAVID HAMBLING
Popular Science, 21 November 2016

Chicago – Small drones are already effective weapons for urban warfare — when armed with miniature warheads, these stealthy spies can turn into lethal assassins. So far their biggest limitation is battery life, but Boeing's patent for a drone battle station sets out to overcome that.

The aerospace giant's 'Vehicle Base Station' resembles Amazon's proposed recharging stations on street lights, but with a different mission. John Vian, a research fellow at Boeing, says the station's main applications are likely to be civil and commercial — used for firefighting and search-and-rescue, for example — but the patent has a decidedly military slant. "The unmanned aerial vehicles may monitor for undesired activity… [which] may be the placement of an improvised explosive device in roadway."

Before coming up with the base station idea, Vian and his colleagues had been looking for the next big thing, and they zeroed-in on 'fractionated systems' — groups of small ground and air robots working together to carry out tasks. "The patent arose from the recognition that these smaller smarter platforms may have limited endurance," Vian told *Popular Science*.

The patent describes a squad or platoon of perhaps 10 to 30 drones, able to follow targets or perch and observe like a fleet of mobile CCTV cameras. When a drone runs low on power it returns to the base station, which physically replaces the drone's batteries. With no wait for recharging, the drone can fly off again at once. The autonomous battery-swapping device would "enable long, 24/7, uninterrupted missions," says Vian. In principle, the base station would allow drones to operate in an area for days, weeks, or months at a stretch.

Clandestine operation is part of the plan. The patent suggests a solar-powered station could be placed on a rooftop where it cannot be seen. Alternatively, the station may be attached to a power line to suck electricity directly from the grid. It could still be covert, hiding in plain sight; few people pay attention to all those black boxes attached to power lines. If some base stations are lost to enemy action, multiple spares would ensure there was always capacity.

Boeing also envisions a line of base stations positioned at intervals of several miles, like gas stations for drones on long journeys. This would allow them to fly right into a combat zone rather than having to be dropped by aircraft.

According to the patent, the base station could act as a shelter in bad weather, a sort of pigeon loft for drones. It may also be a communication hub, housing a satellite microwave link to a remote operations center. In one version of the design, the drones fly missions autonomously, uploading the data they have gathered and receiving fresh instructions each time they return. This would allow them to operate even if all radio signals were being jammed, a growing counter-drone tactic.

The base station might also be able to reload drones with a 'dispensable payload.' Fire suppressant is mentioned as an example, though this seems optimistic for small drones — bombs are a more obvious choice. Hezbollah recently released a video of small drones apparently dropping grenades on enemy positions. Similarly, the US's Switchblade, a lethal drone weighing less than five pounds, carries out kamikaze-style attacks via its on-board munitions. But a reusable attack drone has obvious advantages, especially when the potentially hazardous return-and-reload operation is carried out remotely.

The developers built an indoor test bed to prototype the base station concept using small drones, and they say translating the system into real life deployments would not be difficult.

"With good engineering design, I don't see major challenges for making this system functional," says Boeing Commercial Airplanes Senior Research Engineer Emad Saad. "Precise relative navigation is a challenge, but can be addressed with modern sensors and good engineering design."

Boeing is one of the dominant players in the military market, producing aircraft like the B-52 bomber, F-15 Eagle, and AH-64 Apache attack helicopter. If the base station moves from concept to reality, it would allow swarms of low-cost drones to carry out missions at long range and over long durations, shaking up the market and maybe the entire face of warfare. Perhaps the next big thing from Boeing will be much smaller than the aircraft we usually associate with the name.

Source: https://www.popsci.com/boeing-has-patented-drone-battle-station

Snowboarding Santa Claus goes viral as he shreds behind human-flying drone

By MARISSA PAYNE
The Washington Post, 23 December 2016

Rovaniemi, Finland – Casey Neistat didn't have time to feel scared when he strapped himself to the drone that was about to fly him off the side of a cliff in Finland.

"I was directing eight camera people," said the 35-year-old YouTube star, whose holiday-themed video of the feat went viral this week. "There were so many responsibilities on set, and there's so much to think about just from, like, a pragmatic perspective when you're actually being flown by one of these things, that there wasn't enough room for that kind of emotion."

It also helped that Neistat had "absolute trust" in the man operating the drone not to fly him into a tree as he snowboarded behind it, merely holding a line.

Justin Oakes is "as good as they get," Neistat said in a phone interview Wednesday, by which point his video had already garnered more than 3 million views. As of this posting on Friday, the video had surpassed 5.5 million views.

Oakes, a former drone racer who now runs Droneworks Studios, said in an email that he's been dreaming of making a drone powerful enough to carry a human "for years" but that it was only in the past year or so that it became clear it would become a reality.

With Samsung footing much of the bill, terms of which were undisclosed, Oakes and his team at his Houston-based company got to work. He said the math alone took about a month to complete before any physical building could take place. There was a lot to think about and even more to test.

"We tested the motors, propellers and batteries individually on a bench for stress and to measure available power. Then we assembled the mock-up frame and tested with 200 pounds of dummy weight," Oakes said. "That was several months before the project filmed."

According to Neistat, the original plan was to fly him over a Caribbean island on a surfboard. He brought the idea up to Samsung, whom he's worked with before, earlier this year, but never expected the company to actually agree to do it.

"It's just one of those things you say that you never actually mean," he said.

When logistics made the island-hopping idea impossible, however, Neistat said the team settled on a holiday-themed stunt. They chose Finland for two reasons.

"It seemed very natural to make a Christmas holiday movie at the only place called Santa's Village," he said, referring to the amusement park in the Lapland region of the country.

The second reason was that it was legal to fly a drone powerful and heavy enough to lift a human. In the United States, for example, it's illegal to fly a drone that weighs over 55 pounds. The drone used to lift Neistat weighed 165 pounds on its own. With Neistat attached with his board and the safety rigging, the amount shot up to 365 pounds. It took three people to operate safely.

"Definitely do not try [this] at home!" Oakes warned others who might be thinking of jerry-rigging their own human-flying drone. “It was certainly not something we threw together as a quick stunt. This was extremely complex for a team of pro builders to do in a proper office with pro tools and a budget to make everything safe. … No corners were cut.”

Neistat said that being competent on a snowboard was also essential to pulling off the feat, especially before lift off. While he was hooked up to safety equipment to film the flying scenes, staying upright was entirely his doing when he was being pulled through the trails of Santa’s Village and up a ski slope up to 35 mph.

“In all the shots where I’m being dragged, I’m actually riding,” he said, reasoning that if he were to have lost his balance, he could have let go of the drone.

“This is one of the more fun or exciting or certainly unique things I’ve done,” said Neistat, who previously snowboarded behind a car in the streets of New York City and played real-life Pokemon Go on a skateboard in two of his other viral videos.

But it’s not the flying off the side of a cliff and dangling some 150 feet in the air from a drone that he found the most thrilling about his latest project.

“The most fun part was being pulled by far, especially that scene where I’m actually on a ski mountain and I’m going straight up the mountain and the snowboarders are coming down,” he said. “If you’ve ever skied, that’s something you’ve always fantasized about. It was like having a 35 mph tow rope pulling you up.”

https://www.washingtonpost.com/news/early-lead/wp/2016/12/23/snowboarding-santa-claus-goes-viral-as-he-shreds-behind-human-flying-drone/?utm_term=.d65b49732b58

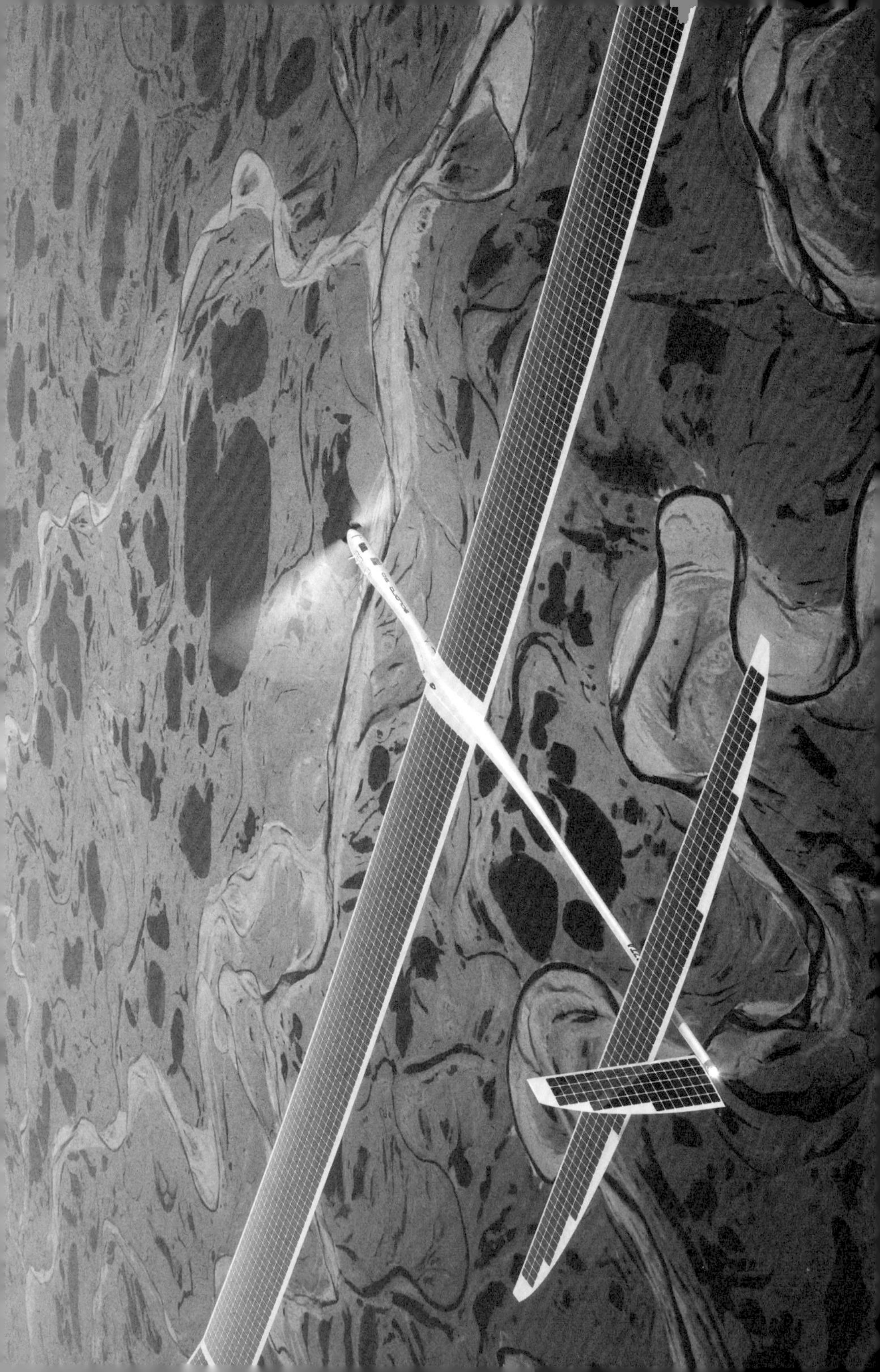

Dutch police are training eagles to remove drones that pose a public safety threat. ©AP/Dutch Police

The 2016 World Drone Prix Dubai aerial track © World Drone Prix

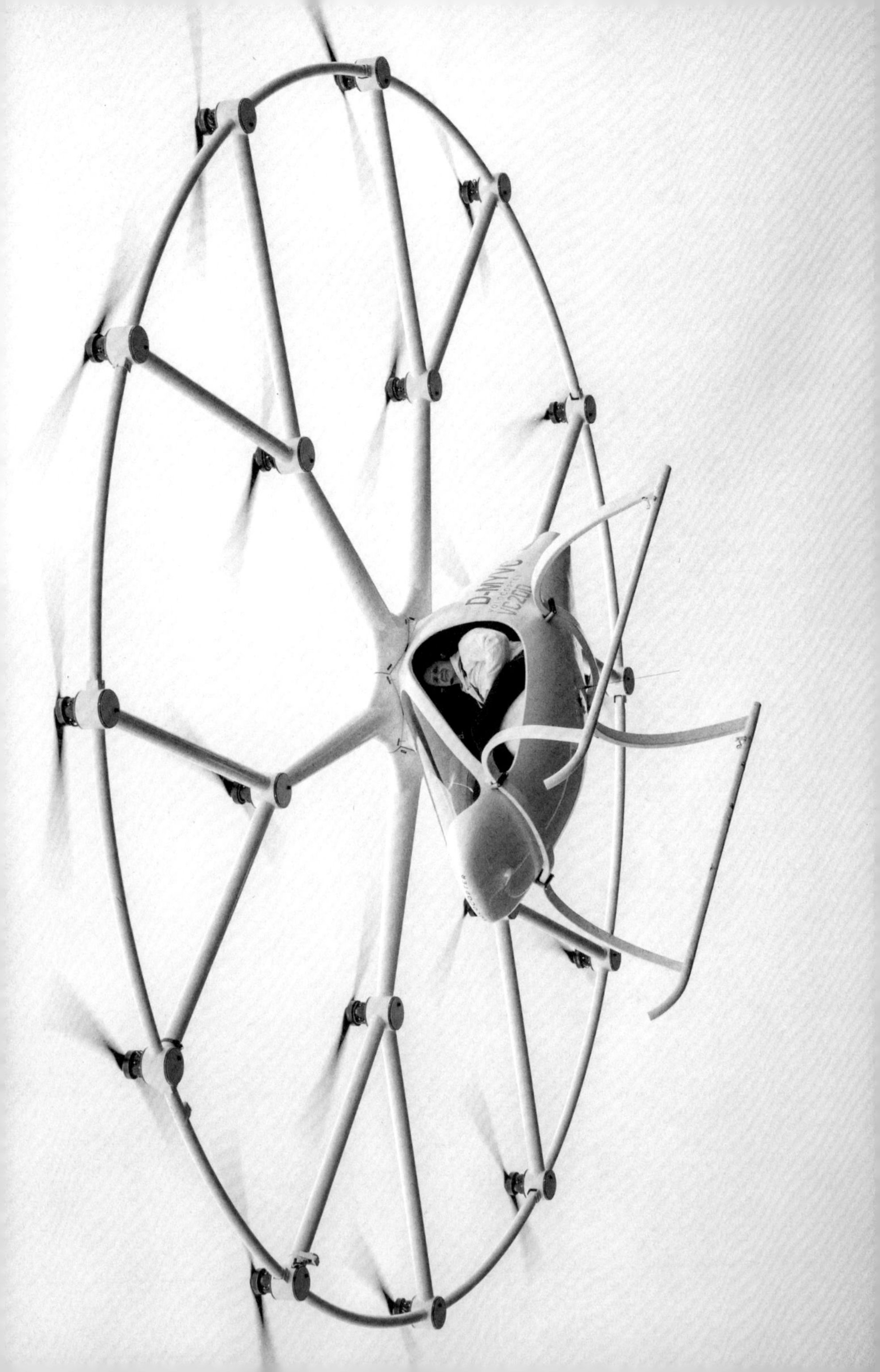
D-MYVC
VC200

An MQ-1B Predator takes off from Ali Base, Iraq. © US Air Force/Tech. Sgt. Sabrina Johnson

Facebook's Aquila drone will be part of a fleet providing internet access to parts of sub-Saharan Africa and beyond. © Facebook

Hussam al-Mayali, a contractor for the Ministry of Interior from Basra, pilots a drone to monitor Isis positions in Fallujah. ©Laith al-Haydar

Snowboarding-Santa Casey Neistat flies over Finland with a custom human-flying drone. ©Youtube/Casey Neistat

A

ABI Activity-Based Intelligence
ADNI Asst. Director of National Intelligence
AF/AFG Afghanistan
AFSOC Air Force Special Operations Command
AFO Advance Force Operations
AFRICOM US Africa Command
ANSF Afghan National Security Forces
AO Area of Operations
AOR Area of Responsibility
AQ Al Qaeda
AQ FAC Al Qaeda Facilitator
AQIM Al Qaeda in the Islamic Maghreb
AQSL Al Qaeda Senior Leadership
AP Arabian Peninsula
APG Aerial Precision Geolocation
ARF Almost Ready to Fly
AUMF Authorization for Use of Military Force

B

BDA Battle Damage Assessment
BDL Bed-Down Location
BVLOS Beyond Visible Line of Sight

C

CAOC Combined Air and Space Operations Center
CAP Combat Air Patrol
CAPE Department of Defense Office of Cost Assessment and Program Evaluation
CCMD Combatant Command
CDE Collateral Damage Environment
CELLEX Cellular Exploitation
CF Coalition Forces
CIA Central Intelligence Agency
CIVCAS Civilian Casualties
CJSOTF Combined Joint Special Operations Task Force
C-LRA Counter-Lord's Resistance Army
CNO Computer Network Operations
CoM Chief of Mission (Ambassador)
COMINT Communications Intelligence
CONOP Concept of Operations
CoS Chief of Station (CIA)
CT Counterterrorism

D

DEPSECDEF Deputy Secretary of Defense
DNR Dial Number Recognition
DoD Department of Defense

E

EKIA Enemy Killed in Action
EWIA Enemy Wounded in Action

F

FAA Federal Aviation Administration
FATA Federally Administered Tribal Area
F3EA Find, Fix, Finish, Exploit/Analyze
FFF Find, Fix, Finish
FMV Full Motion Video
FOB Forward Operating Base
FOS Flight Operation Supervisor
FPV First Person View
FVEY Five Eyes

G

GCC Geographic Combatant Command
GCS Ground Control Station
GIS Geographic Information System
GOCO Government-Owned Contractor-Operated
GPS Global Positioning System
GSM Global System for Mobile Communication

H

HN Host Nation
HOA Horn of Africa
HUMINT/HI Human Intelligence
HVI High Value Individual

I

IC Intelligence Community
ICAO International Civil Aviation Organization
IED Improvised Explosive Device
IIR Intelligence Information Report
IMINT Imagery Intelligence
ISAF International Security Assistance Force
IPB Intelligence Preparation of the Battlespace
IPOE Intelligence Preparation of the Operational Environment
ISR Intelligence, Surveillance, and Reconnaissance

J

JARUS Joint Authorities for Rulemaking on Unmanned Systems
J2 Intelligence Directorate of a Joint Staff
J5 Plans Directorate of a Joint Staff
JLENS Joint Land Attack Cruise Missile Defense Elevated Netted Sensor System
JP Jackpot

JTL	Joint Target List
JWICS	Joint Worldwide Intelligence Communications System
JOC	Joint Operations Center
JPEL	Joint Prioritized Effects List
JSOC	Joint Special Operations Command

L

LOCUST	Low-Cost Unmanned aerial vehicle Swarming Technology

M

MAM	Military-Age Male
MARSOC	Marine Corps Special Operations Command
MEDEX	Media Exploitation
MIC	Mission Intelligence Coordinator
mIRC	Military Intelligence Reserve Command
MFW	Medium Fixed-Wing
MON	Memorandum of Notification
MTOW	Maximum Take-off Weight
MTS	Masked Target Sensor
MQ-1	General Atomics Predator
MQ-9	General Atomics Reaper

N

NAI	Named Area of Interest
NAVAF	US Naval Forces Africa
NSA	National Security Agency
NSB	National Security Bureau

O

OEF	Operation Enduring Freedom
OIF	Operation Iraqi Freedom
OBJ	Objective
ODTAAC	Outside a Defined Theater of Active Armed Conflict
OPTEMPO	Operational Tempo
OPSEC	Operations Security
OSD	Office of Secretary of Defense

P

PACOM	US Pacific Command
PAX	Passengers
PID	Positive Identification
PTT COMINT	Push-to-Talk Communications Intelligence
PNF	Plug and Fly
POL	Pattern of Life
POTUS	President of the United States

R

RAF	Royal Air Force
ROE	Rules of Engagement
ROYG	Republic of Yemen Government
RPA	Remotely Piloted Aircraft
RPAS	Remotely Piloted Aircraft System
RTF	Ready To Fly

S

SA	Situation Assessment; Surface-to-Air; Security Assistance; Special Agent; System Administrator
SAR	Search and Rescue
SECDEF	Secretary of Defense
SIGINT/SI	Signals Intelligence
SITREP	Situation Report
SME	Subject Matter Expert
SNA	Social Network Analysis
SOCOM	US Special Operations Command
SO	Sensor Operator
SOTF	Special Operations Task Force
SRA	System and Resource Analyses
SSW	Small Smart Weapon

T

TADS	Terror Attack Disruption Strikes
TB	Taliban
TF	Task Force
TIR	Tactical Interrogation Report
TTP	Tactics, Techniques, Procedures

U

UAV	Unmanned Aerial Vehicle
UCAV	Unmanned Combat Aerial Vehicle
USAF	United States Air Force

V

VBIED	Vehicle-Borne Improvised Explosive Device
VEO	Violent Extremist Organization
VID	Voice ID
VLOS	Visual Line of Sight
VTOL	Vertical Take Off and Landing

W

WMD	Weapon of Mass Destruction

THE DRONE CHRONICLES

Concept and design: Rob van Leijsen
Authors: Joerg Bader, Nicolas Nova, Brice Pauset, Rob van Leijsen
Publisher: Joerg Bader
Distribution: Spector Books, Leipzig / les presses du réel, Dijon
Editorial coordination: Tiphaine Carrère, Olivia Fahmy, Maud Pollien, Natalia Reichert, Alexandra Theiler
Translation: Judith Hayward, Natalia Reichert
Copy-editing and proofreading: Natalia Reichert

Printing and binding: Musumeci S.p.A., Quart
ISBN: 978-3-95905-310-5

Thanks to:
Markus Farner, Cintia Stucker, Veronica Tracchia

éditions Centre de la photographie Genève
www.centrephotogeneve.ch

Spector Books
www.spectorbooks.com

The Centre de la photographie Genève is supported by Ville de Genève

With the support of:

VILLE DE NYON · Région de Nyon · FONDATION LEENAARDS